WASTE CONTAINMENT SYSTEMS:

Construction, Regulation, and Performance

Proceedings of a Symposium sponsored by the
Committee on Soil Improvement and Geosynthetics
and the Committee on Soil Properties of the
Geotechnical Engineering Division,
American Society of Civil Engineers
in conjunction with the
ASCE National Convention
San Francisco, California
November 6-7, 1990

GEOTECHNICAL SPECIAL PUBLICATION NO. 26

Edited by Rudolph Bonaparte

Published by the
American Society of Civil Engineers
345 East 47th Street
New York, New York 10017-2398

ABSTRACT

This proceedings consists of papers presented at a symposium on waste containment systems at the ASCE Convention in San Francisco, California on November 6th and 7th, 1990. Geotechnical engineers are increasingly involved in the design and construction of waste containment systems that utilize soil and geosynthetic liners and drainage layers. This book presents twelve papers on the construction, regulation, and performance of waste containment systems. The papers cover topics such as: 1) federal and state landfill containment regulations; 2) performance evaluations of earthen liners; 3) field behavior of double-liner systems; 4) long-term properties of earthen and geomembrane liners; and 5) detailed case studies of earthen liner performance. It also discusses construction quality control of earthen and geomembrane liners, field verification of earthen liner hydraulic conductivity, and the use of waste-attenuating soil materials, such as bentonites, zeolites, organically modified clays, and fly ash, in earthen liners.

Library of Congress Cataloging-in-Publication Data

Waste containment systems: construction, regulation, and performance: proceedings of a symposium/sponsored by the Committee on Soil Improvement and Geosynthetics and the Committee on Soil Properties of the Geotechnical Engineering Division, American Society of Civil Engineers in conjunction with the ASCE National Convention, San Francisco, California, November 6-7, 1990; edited by Rudolph Bonaparte.

p. cm.

Includes indexes.

ISBN 0-87262-787-X

1. Sanitary landfills—Design and construction—Congresses. 2. Engineering geology—Congresses. I. Bonaparte, Rudolph. II. American Society of Civil Engineers. Committee on Soil Improvement and Geosynthetics. III. American Society of Civil Engineers. Geotechnical Engineering Division, Soil Properties Committee. IV. ASCE National Convention (1990: San Francisco, Calif.)

TD795.7.W37 1990

628.4'4564—dc20 90-49336

CIP

Library of Congress Catalog Card No: 90-49336

ISBN 0-87262-787-X

Manufactured in the United States of America.

GEOTECHNICAL SPECIAL PUBLICATIONS

1) TERZAGHI LECTURES
2) GEOTECHNICAL ASPECTS OF STIFF AND HARD CLAYS
3) LANDSLIDE DAMS: PROCESSES RISK, AND MITIGATION
4) TIEBACKS FOR BULKHEADS
5) SETTLEMENT OF SHALLOW FOUNDATION AND COHESIONLESS SOILS: DESIGN AND PERFORMANCE
6) USE OF IN SITU TESTS IN GEOTECHNICAL ENGINEERING
7) TIMBER BULKHEADS
8) FOUNDATIONS FOR TRANSMISSION LINE TOWERS
9) FOUNDATIONS AND EXCAVATIONS IN DECOMPOSED ROCK OF THE PIEDMONT PROVINCE
10) ENGINEERING ASPECTS OF SOIL EROSION DISPERSIVE CLAYS AND LOESS
11) DYNAMIC RESPONSE OF PILE FOUNDATIONS—EXPERIMENT, ANALYSIS AND OBSERVATION
12) SOIL IMPROVEMENT—A TEN YEAR UPDATE
13) GEOTECHNICAL PRACTICE FOR SOLID WASTE DISPOSAL '87
14) GEOTECHNICAL ASPECTS OF KARST TERRAINS
15) MEASURED PERFORMANCE SHALLOW FOUNDATIONS
16) SPECIAL TOPICS IN FOUNDATIONS
17) SOIL PROPERTIES EVALUATION FROM CENTRIFUGAL MODELS
18) GEOSYNTHETICS FOR SOIL IMPROVEMENT
19) MINE INDUCED SUBSIDENCE: EFFECTS ON ENGINEERED STRUCTURES
20) EARTHQUAKE ENGINEERING & SOIL DYNAMICS (II)
21) HYDRAULIC FILL STRUCTURES
22) FOUNDATION ENGINEERING: CURRENT PRINCIPLES AND PRACTICES
23) PREDICTED AND OBSERVED AXIAL BEHAVIOR OF PILES
24) RESILIENT MODULI OF SOILS: LABORATORY CONDITIONS
25) DESIGN AND PERFORMANCE OF EARTH RETAINING STRUCTURES
26) WASTE CONTAINMENT SYSTEMS: CONSTRUCTION, REGULATION, AND PERFORMANCE

PREFACE

Geotechnical engineers are increasingly involved in the design and construction of waste containment systems that utilize soil and geosynthetic liners and drainage layers. The role of geotechnical engineers in the field of waste containment has been recognized by the Geotechnical Engineering Division for some time. In 1977, the Division sponsored the conference on *Geotechnical Practice for Disposal of Solid Waste Materials* and in 1987 they sponsored the conference on *Geotechnical Practice for Waste Disposal '87*. Since 1987, the field has evolved significantly, due to the promulgation of ever more stringent federal and state regulations, and to a rapidly increasing understanding of the properties and capabilities of the materials used in construction. Accordingly, the Committee on Soil Improvement and Geosynthetics and the Committee on Soil Properties proposed a three-session symposium on recent developments in construction, regulation, and performance of waste containment systems. The November 1990 ASCE National Convention in San Francisco, California was selected as the venue for the symposium.

All of the papers presented in this symposium were invited by the sponsoring committees. The papers, although invited, were peer-reviewed for technical quality and content before being accepted. The standards of review were the same as for the ASCE *Journal of Geotechnical Engineering*. Each paper received two positive reviews before being accepted and was revised to conform to any mandatory revisions of the reviewers. It should be noted that all papers in this volume are eligible for discussion in the *Journal of Geotechnical Engineering* and for ASCE awards.

This symposium was organized under an accelerated timeframe and would not have been possible without the significant support of Loren R. Anderson, Chairman of the Committee on Soil Improvement and Geosynthetics, Gary M. Norris, Chairman of the Committee on Soil Properties, and J. Michael Duncan, Chairman of the Committee on Sessions. The symposium organizers also owe their gratitude to the following individuals who peer-reviewed the symposium papers, also under an accelerated timeframe.

Loren R. Anderson
Robert C. Bachus
Richard D. Barksdale
John F. Beech
R. Jeffrey Dunn
Robert D. Holtz
Sandra L. Houston
Keith A. Kessler
Robert M. Koerner
Paul C. Knodel
Robert Y. Liang
Scott M. Luettich
Namunu J. Meegoda
George A. Munfakh
Gary M. Norris
Bertrand S. Palmer
Richard P. Ray
Gregory N. Richardson
Glenn J. Rix
Charles D. Shackelford
Kevan D. Sharp
C. Joel Sprague
Timothy D. Stark
Richard L. Wiltshire

Finally, the symposium organizers extend their thanks to Shiela Menaker and the staff at ASCE who managed the assembly and printing of this volume.

Rudolph Bonaparte
Proceedings Editor

CONTENTS

Session 1: Regulation and Performance

Session 2: Long-Term Performance

Session 3: Construction, Quality Control, and Future Directions

LANDFILL CONTAINMENT SYSTEMS REGULATIONS

Robert E. Landreth[1]

The Congress of the United States, through the Resource Conservation and Recovery Act of 1976 (RCRA) and its 1984 Hazardous and Solid Waste Amendments (HSWA), mandated the United States Environmental Protection Agency (EPA) to develop standards for the management of both hazardous and nonhazardous wastes in this country in a manner that would protect human health and the environment. In response, the EPA through its experience in field activities and research developed interim guidelines. Many of these guidelines have evolved into current EPA regulations, and the Agency continues to refine the control strategies. The regulations and guidance documents will continue to be updated as construction materials are improved and as design concepts are better understood and verified in the field.

The current status of hazardous waste regulations (Subtitle "C" of RCRA as amended by HSWA) and nonhazardous waste regulations (Subtitle "D") including proposed regulations are summarized below. The role of geosynthetics, including flexible membrane liners (FMLs), geonets, geotextiles, and plastic pipes, in meeting the regulatory requirements will be discussed. Cover systems and liner systems are discussed rather than the individual components.

SUBTITLE "C" REGULATIONS

Under Section 3004 of RCRA, owners and operators of treatment, storage, and disposal facilities (TSDFs) are required to comply with standards "necessary to protect

---

[1]Chief, Municipal Solid Waste and Residuals Management Branch, Waste Minimization, Destruction, and Disposal Research Division, Risk Reduction Engineering Laboratory, U.S. Environmental Protection Agency, Cincinnati, OH 45268.

human health and the environment." Since enactment of RCRA, EPA has promulgated interim status standards (40 CFR Part 265) and permitting standards (Part 264) governing the design, operation, and maintenance of landfills, surface impoundments, and waste piles used to treat, store, or dispose of hazardous wastes (Federal Register, 1986a and 1987b).

In the fourteen years since the passage of RCRA, the standards have been modified to incorporate new information developed through research and field experience. Most recently the Agency has also proposed new performance and design standards for control of leaks, including leak detection, the establishment of an action leakage rate, and a permit requirement for inclusion of a response action plan (Hokanson, 1988). The current schedule for publishing these final regulation is the fall of 1990. Although existing facilities will not have to meet these regulations, existing double-liner facilities may be required to develop action leakage rate and response action plan criteria based on site-specific capabilities.

When the Hazardous and Solid Waste Amendments (HSWA) were passed in November 8, 1984 the liner containment requirement for landfills was for two or more liners with a leachate collection systems above and between the liners. To clarify the HSWA a rule issued on July 15, 1985 set top liner standards that could be met by a flexible membrane liner (FML) and a bottom liner that could be met by three feet of compacted soil or other natural materials with a hydraulic conductivity equal to or less than $1X10^{-7}$ cm/sec. The rule was reviewed and modeled in saturated and unsaturated hydraulic flow conditions. The result of these studies is the current Agency policy for hazardous waste landfills requiring a double liner which has a composite bottom liner and a top FML, Figure 1. The composite bottom liner is one that consists of a FML on a compacted, low-permeability, natural soil.

LINER SYSTEMS

The liner system currently being proposed by most hazardous waste management facilities incorporate in descending order a filter layer, followed by a primary leachate collection and removal system (LCRS), a primary FML, a leak detection, collection and removal system (LCDRS), and a composite liner above the native soil foundation (EPA, 1987). The composite liner is defined as a FML an a compacted, low permeability ($k \leq 1X10^{-7}$ cm/sec) natural soil.

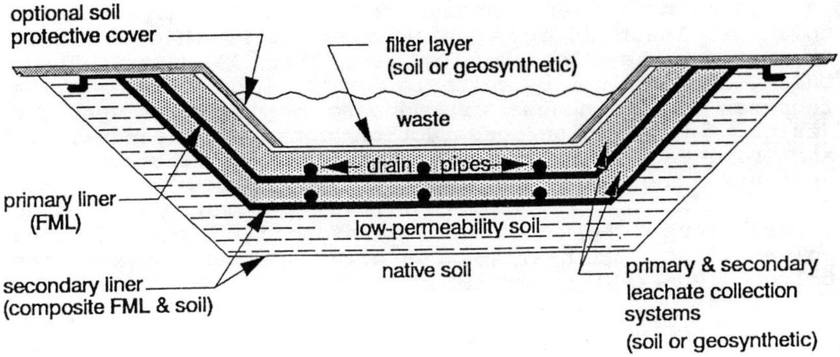

Figure 1. Schematic of Double Liner System.

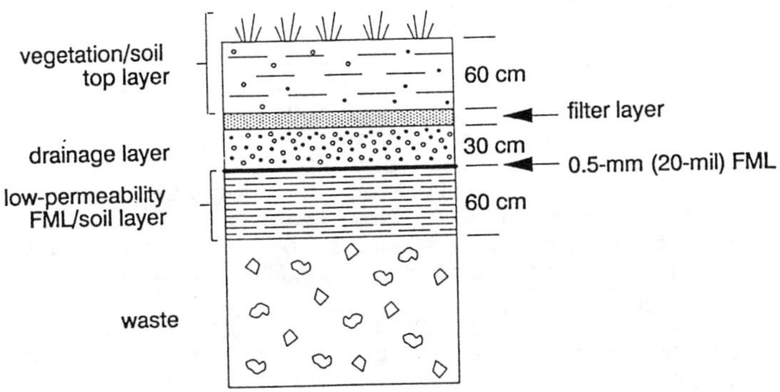

Figure 2. Schematic of Cover System.

In bottom liner systems, regulations require the FML to be at least 30 mils thick or 45 mils thick if left exposed to the elements for more than 30 days. These thicknesses may not be suitable for all FML materials. The required FML thickness will depend on the site-specific design, installation/construction concerns, seamability, and long-term durability. Taking these factors into account, the EPA recommends the following minimum thicknesses: polyvinylchloride (PVC) - 30 mil, chlorinated polyethylene (CPE) - 30 mil, reinforced chlorosulfonated polyethylene (CSPE-R) - 36 mil and semi-crystalline polyethylene (PE) - 60 mil.

Chemical Resistance

Chemical compatibility or chemical resistance to the waste or leachate to be contained has been and continues to be a major factor in selecting the FML for a waste containment facility. The increased use of other types of geosynthetics has raised similar issues. The Agency has a long standing policy (EPA, 1966; Matrecon, 1988) that all materials coming in contact with leachate must be chemically resistant to that leachate. The Agency uses the EPA Method 9090 as the immersion procedure to expose the candidate materials to the leachate for various periods of time. The procedure requires that samples of the geosynthetic be immersed in the representative sample of leachate from the waste management unit for a periods of 30, 60, 90 and 120 days at temperatures of 23°C and 50°C. After each exposure period, various physical property tests are performed to determine changes in these properties. The containing vessel should not be of the same material as that being tested and should not compete with the geosynthetics being evaluated for potentially aggressive leachate constituents. The leachate vessel should be sealed with no free air space in order to prevent the loss of volatile constituents from the leachate.

An alternative immersion procedure is being developed by the American Society for Testing and Materials (ASTM) Committee D-35 Committee on Geosynthetics. This ASTM procedure closely follows the procedure of EPA Method 9090 but adds details regarding test conditions and the immersion vessel. This procedure is under review by the EPA for potential acceptance in lieu of portions of Method 9090.

As a part of the Method 9090 test, procedures are identified for evaluating the physical properties of FML, but other types of geosynthetics were being used in waste management facilities and a consistent set of physical property evaluation tests were needed. Therefore, Agency in consultation with several research and testing firms developed initial recommendations (Landreth, 1990) for evaluating the chemical resistance of geosynthetics. This paper identifies ASTM standard tests for geotextiles, geonets, geogrids and plastic pipe.

The paper also discusses the need for finger-printing geosynthetics. Fingerprinting characterize the material in general terms such as polymer type and amount that is used; the amount of extractables; and the carbon black and ash contents. Fingerprinting, is performed to help the Agency determine that the geosynthetic material evaluated for chemical resistance in laboratory testing is essentially identical to the geosynthetic material installed in the waste management unit.

Chemical resistance must be achieved if the geosynthetic material is to survive. The Agency has long recognized that no single material is resistant to all wastes and has never granted blanket approval to any generic type of geosynthetic. Recognizing the importance of chemical resistance the Agency initiated several modifications to the type of waste that could be disposed of in landfills. The EPA has banned the placement of certain classes of chemicals and liquids in landfills has improved the potential "quality" of any leachate that may be generated. These modifications should increase the number of FML compositions available for use. The increased number of FML compositions should now allow the designer to develop innovative designs. We believe innovative designs will be more economical, technically viable, and be more reliable.

DESIGN ELEMENTS

The actual design of the landfill lining system considers not only the chemical resistance of the geosynthetic but the minimum technology guidance (EPA, 1987), material stress considerations, structural details, and panel fabrication. The development of a comprehensive Quality Assurance/Quality Control program will help ensure that the landfill design is constructed to specifications and with specified materials.

Technology Guidance: To achieve EPA guidelines for minimum leakage, the liner should not leak at a rate greater than the action leakage rate stated in the site's response action plan as determined by the owner/operator and the permit writer. The minimum thickness of FML is set at 0.75mm (30 mils) or 1.2mm (45 mils) if it will be exposed for more than 30 days. State requirements may be more stringent than EPA guidelines; the more stringent requirements must be followed in design. It should be noted increasing the liner thickness can greatly increase the cost. Increasing the liner thickness of the same material from 0.75mm to 1.52mm (30 to 60 mils) can double the per acre cost of the FML, the actual increase will depend on the polymer type.

Stress Considerations: Stress conditions must be considered for both the bottom and the side slopes of a landfill. For side slopes, the weight of the FML itself and the weight of the waste must be reckoned with. Because the geosynthetic must be able to support its own weight on the side slopes, the specific gravity, thickness, and yield stress of the material and the friction angle must be known to calculate self-weight. For exposed FMLs, uplift forces caused by wind is another critical loading condition. For FMLs covered with a layer of soil, the tendency of the soil to slide down the slope is a critical leading conditions and should be evaluated. For the bottom two different mechanisms by which loads can be applied to the FML should be evaluated. The first mechanism is due to compression of consolidation of the foundation soil supporting the landfill. The second mechanism is due to compression or consolidation of the waste.

Structural Details: Anchorage, access ramps, collection standpipes, and penetrations are all structural details that warrant attention, especially in double liner systems.

In anchorage trenches, the geosynthetic can be ripped or pulled out. The pullout capacity for various anchorage configurations (e.g., trenches with box or V-shape) can be calculated (Koerner, 1990). Research is currently being conducted at Drexel University's Geosynthetic Research Institute to evaluate this design approach.

Most facilities have access ramps to bring waste into the landfill. With double liner systems, the continuity of both liners must be maintained over the entire surface. Construction activities such as traffic induced damage and site drainage must be addressed in the design. Vertical collection standpipes are used to access the primary leachate collection sumps. As waste settles the standpipe can be affected by downdrag and the primary FML beneath can be punctured. ASTM-D-2435 is a one dimensional consolidation test that could be used to measure the consolidation properties of the waste. From these properties, the settlement of the waste can be calculated. This calculated settlement is used to calculate the downdrag forces acting on the stand pipe. The settlement of the standpipe due to the downdrag forces can them be calculated to determine the impact on the FML. Remedies could include coating the standpipe with a viscous or low-friction coating or encapsulating it with multiple layers of low friction resistance material. A flexible foundations or spread footer design could provide a more gradual transition and spread the distribution of contact pressures over a larger portion of the FML than would a rigid foundation design.

The leak detection, collection and removal system should be designed with side slope risers to avoid penetration of the primary FML.

Panel Layout: The layout of FMLs in the landfill should be planned so that seams run up and down the slope and that the length of field seams is minimized whenever possible.

LEACHATE COLLECTION SYSTEMS

During the active life of a waste management unit and, to a lesser extent, after closure, some of the rainfall falling on the unit percolates into and through the waste. This liquid, called leachate, contains chemical constituents leached from the waste. Leachate is both site-specific and waste-specific with regard to its quantity and quality. To prevent the accumulation of leachate in the bottom of landfill facilities, collection and removal systems are constructed as shown in Figure 1, to facilitate the removal of liquids for treatment. There are at least two leachate collection systems, the primary leachate collection and removal system (LCRS) and the leak detection, collection, and removal system (LDCRS), in a modern waste management facility.

The leachate will first contact the LCRS located directly below the waste and above the top liner. The design and construction of this system is based on site-specific conditions. LDCRS is between the top and composite liners. Ideally, the LDCRS system should see only small amounts, if any, of leachate, but it is usually designed to accommodate total failure of the primary liner and collection system.

Materials of Construction

Until recently, collection systems were constructed exclusively of natural materials, i.e., sands and gravels. However, the substitution of geosynthetics for the natural materials has given engineers greater freedom in developing innovative designs to efficiently collect and remove excess liquids. Geosynthetics used in collection systems include geotextiles, for filtration and for separation of structural components; geonets for in-plane liquid movement; and plastic pipes for collection and removal. Natural materials may still be used either independently or in combination with geosynthetics.

Geotextile filters are designed (Carroll, 1987; FHWA, 1987; Giroud, 1982; Koerner, 1990) to prevent fine soil particles from entering and fouling the drainage media. Although filter designs using sands and gravels have been commonly used, they take up valuable space and may move under loading conditions, especially on slopes. The open spaces in a geotextile are designed to allow liquid flow perpendicular to the surface without becoming plugged with soil. Geotextiles have the advantage of requiring less vertical space than granular soil filters, resulting in more space for waste containment. Geotextiles are also easy to install and to anchor against slope instability. They have the disadvantage of clogging with soil if not properly designed. An ongoing study (Koerner, 1989) indicates that geotextiles can also become biologically clogged very quickly in municipal solid waste landfill leachate collection systems. This study is also investigating methods, including the use of biocides, to prevent biological clogging.

To meet high in-plane flow rate (transmissivity) criteria, a geonet may be used. Geonets require less space than granular soil for equivalent flow capacity, thus promoting rapid transmission of liquids. Because of larger apertures, they are less likely to clog than soils or geotextiles. Depending on the overall design, they may require geotextile filters on one or both sides to prevent the entrance of soil or waste that would reduce their efficiency. Potential disadvantages over long periods of time and under high compressive loads are creep (a tendency to flatten) and intrusion (material working its way into the openings). Again proper design should minimize the potential for these problems.

Geocomposites, which may combine the features of geotextiles and geonets, have generally not been used in leachate collection systems due to their relatively low crush strength (Koerner, 1990). They may, however, be useful in surface water collector systems, where the applied normal stresses are low.

Plastic pipes are used to collect the leachate and rapidly remove the liquid. Plastic pipes are designed to perform with both natural and geosynthetic drainage media.

Design Considerations

The design of leachate collection systems is based on hydraulic flow in the media, the chemical resistance of the media and the strength of the media in resisting overburden pressures. Since collection system designs are a function of the amount of the liquid they may have to transport, it is necessary to determine the maximum removal rate that may be required at a given site.

The Hydrologic Evaluation Landfill Performance (HELP) computer model can be used to estimate the amount of leachate that may travel through the system (EPA, 1984) HELP was developed to assist in estimating the magnitude of water-balance components and the height of water-saturated soil above the barrier layer. The program has historical data for 184 cities throughout the United States and other soil and vegetation data that can be used in a default mode in the event local data are unavailable. A simple water balance method may also be used to calculate leachate volumes.

Once the rate of leachate generation has been determined, the hydraulic flow capacity, for a granular drainage layer needed to handle that generation rate, must be calculated. The flow capacity is dependent on certain physical soil characteristics of the drainage layer. EPA's Minimum Technology Guidance recommends that granular drainage layers have the following minimum characteristics:

- be 30 cm (12 in) thick
- 1.00 cm/sec (2.0 ft/min) minimum hydraulic conductivity
- a slope $\geq$ 2%
- a perforated pipe and a layer of filter soil, and
- cover the bottom and side wall of the facility.

Collection pipe hydraulic designs are usually taken directly from nomographs developed for the specific pipe used. The overburden pressure on the pipe during the active life will cause the pipe to deform. A normal limiting deflection value of twenty percent of the pipe diameter is often used for plastic pipe. This is compared with the trench geometry to determine if the deflection value is acceptable.

Geosynthetic drainage materials or geonets are produced by a variety of manufacturers. Their physical and hydraulic characteristics are published. The ASTM D-4716 test method is generally used to determine the flow rate/transmissivity of the geosynthetic drainage material for a design. The proposed cross section of the design should be modeled as closely as possible, i.e., same profile layer combination. The Agency recommends applying 2 to 3 times maximum overburden pressure during the test to try to account for long-term creep and intrusion of the profile. The planar flow, at a constant hydraulic head, is applied and the flow rate as a function of time through the geonet is measured.

A very important part of a leachate collection system is the removal component. Low-volume submersible pumps can present problems e.g., burn out from constantly running, if the EPA recommended 1-foot head maximum is maintained. Consideration must also be given to whether the system's exit will be vertical, up the side wall, or out through the side. In all cases the design should address the need for periodic clean-out of the system.

COVER SYSTEMS

After a landfill or impoundment has been filled, the regulations require that a multi-layered cover system figure 2, be designed to isolate the waste from the environment, minimize infiltration of surface water, and thus, minimize liquid migration and leachate formation. The cover system must function with minimum maintenance, promote drainage, accommodate settling and subsidence, and have a permeability no greater than that of the liner system.

Generally a vegetated upper layer prevents erosion and promotes evapotranspiration; an underlying drainage layer conveys percolation out of this cover; and a moisture barrier (e.g., FML) prevents infiltration to the waste.

Because site conditions differ so greatly, no one universal design should be used. Rather, a state-registered, qualified design engineer must perform a technical analysis of the overall design to ensure that the performance standards have been achieved for that particular geological setting.

The waste and its placement must be characterized because the cover settlement primarily depends on how the waste mass consolidates, compresses, and collapses. Will the waste settle unevenly? Models have been developed to help estimate the effects of settlement, and covers can be designed to compensate for anticipated settlement (EPA, 1987). The designer must also analyze for slope stability and determine and allow for possible soil erosion.

EPA guidance document (EPA, 1989) recommends that a cover system include first a foundation layer between the waste and the cover, followed, in turn, by a low-permeable soil layer, a geomembrane (min 20 mil thick), a drainage layer, a graduated granular soil or geotextile layer, and finally a vegetated top soil layer, with special attention given to the outer edges of the cover system.

For each of these layers comprising the cover, the materials must be suitable to their task, and a Quality Assurance/Quality Control program for the materials, the construction, the layer thickness, the final grades, etc., must be developed, followed, and documented. The earthwork for cover construction is similar to that used for roadways, although the contact pressure of the equipment must be within an allowable load-bearing range. Preventive maintenance work should be done during the post-closure period to prevent loss of vegetation and gully development.

SUMMARY

The RCRA hazardous waste containment regulations prescribe performance standards rather than design standards or recommendations. The EPA has compiled source documents to aid the designer, permit reviewer, and owner/operator with each element of the design, construction, and what is needed for the approval process. Geosynthetic materials are being increasingly used in the design and construction of containment systems. Geomembranes are required as the mainstay of the low-permeability barriers in liners and covers. Geonets may be substituted for high-permeability soil in drainage layers in the cover and between the liners in a double-liner system. Geotextiles are frequently used to separate layers to prevent material encroachment from one layer into another or to prevent abrasion damage. Plastic pipes are used to quickly carry off liquid from drainage layers. Geosynthetic materials must be carefully selected, tested, and installed to assure that they will carry out their intended functions indefinitely. Otherwise, they are subject to a variety of deterioration mechanisms, such as chemical dissolution, creep under pressure, or rupture under physical stresses. All of these can be prevented by materials testing and careful attention to detail in design, construction, and construction quality assurance.

REFERENCES

Carroll, Jr., R.G. 1987. "Hydraulic Properties of Geotextiles," Geotextile Testing and the Design Engineers, ASTM 952, J.E. Fluet, Jr. Ed., American Society for Testing and Materials, Philadelphia,, 1987, pp. 7-20.

Environmental Protection Agency. 1966. "Supplementary Guidance on Determining Liner/Leachate Collection System Compatibility." August 7, 1966, OSWER Directive 9480.0013.

Environmental Protection Agency. 1984. The Hydrological Evaluation of Landfill Performance (HELP) Model. Vol. 1 and 2, User's Guide for Version 1. EPA/530-SW-84-009 and EPA/530-SW-84-010. NTIS PB 85-200840 and PB 85-100832. Municipal Environmental Research Laboratory, Cincinnati, OH

Environmental Protection Agency. 1987. Minimum Technology Guidance on Double Liner Systems for Landfills and Surface Impoundments -- Design, Construction, and Operation. EPA/530-SW-85-014, Office of Solid Waste. Washington, DC.

Environmental Protection Agency. 1987. Prediction/Mitigation of Subsidence Damage to Hazardous Waste Landfill Covers. EPA/600/2-87/025, NTIS PB 87-175378, Hazardous Waste Engineering Research Laboratory, Cincinnati, OH.

Environmental Protection Agency. 1989. Technical Guidance Document: Final Covers on Hazardous Waste Landfills and Surface Impoundments. EPA-530/SW-89-047 NTIS PB 89-2334-80/AS U.S. Environmental Protection Agency, Cincinnati, OH.

Federal Register, Vol. 51, No. 60, March 28, 1986, pp. 10706-10723.

Federal Register, Vol. 52, No. 103, May 29, 1987, pp. 20218-20311.

"FHWA Geotextile Engineering Manual" Contract No. DTFH61-83-C-00150, March 1985, 917 pg.

Giroud, J.P. 1982. "Filter Criteria for Geotextiles," Proceedings, 2nd International Conference on Geotextiles, Las Vegas.

Hokanson, S., and J. Rubinson. 1988. Liners and Leachate Collections and Removal Systems in Double-Lined Land Disposal Units: Future Design and Performance Requirements. Presented at WSWMA Wastech `88, October 26-27, 1988, Boston, Massachusetts.

Koerner, G. R. and R. M. Koerner. 1989. Biological Clogging of Geotextiles Used as Landfill Fillers, First Year's Results, June 1989 unpublished preliminary report.

Koerner, R. M. 1990. Designing with Geosynthetics - Second Edition. Prentice-Hall.

Landreth, Robert E. 1990. Chemical Resistance Evaluation of Geosynthetics used in Waste Management Applications. Presented at ASTM Symposium on Geosynthetic Testing for Waste Containment Applications, January 23, 1990 Las Vegas, Nevada.

Matrecon, Inc. 1988. Lining of Waste Containment and Other Impoundment Facilities. EPA 600/2-88-052. NTIS PB 89-129670 (SW870, second revised edition.) U.S. Environmental Protection Agency, Cincinnati, OH

REGULATION, CONSTRUCTION AND PERFORMANCE OF CLAY-LINED LANDFILLS IN WISCONSIN

By: Mark E. Gordon,[1] M. ASCE, Paul M. Huebner,[2] and Gene R. Mitchell[3]

ABSTRACT: The Wisconsin Department of Natural Resources (WDNR) has required the use of thick clay liners combined with a leachate collection system as a means of groundwater protection at municipal and non-hazardous industrial solid waste landfills for almost 15 years. During this period, approximately 85 solid waste disposal facilities have been approved using this design strategy. Operational experiences along with data generated from environmental monitoring of many of these facilities have provided a substantial amount of information regarding the performance of clay-lined landfills.

WDNR routinely evaluates available performance data to determine if clay-lined landfills are functioning properly and providing acceptable environmental protection. To date, the data show that use of thick clay liners in conjunction with an efficient leachate collection system is providing a high level of groundwater protection at Wisconsin landfills.

This paper provides background information on the history of Wisconsin's solid waste regulatory program including the mechanisms used to require clay liners and leachate collection systems. Information is also presented on specifications for clay liners, design concepts, and the comprehensive performance monitoring required for clay-lined landfills in Wisconsin. Actual data and experience from Wisconsin facilities are used to supplement the discussions.

INTRODUCTION

Wisconsin's solid waste regulatory program has been in-place for nearly 20 years. The goals in the early years of the solid waste management program were modest. All existing solid waste facilities were identified and licensed. Approximately 2000 landfills existed in the state at that time.

---

[1]Environmental Engineer, Wis. Department of Natural Resources, P.O. Box 7921, Madison, WI 53707

[2]Hydrogeologist, Wis. Department of Natural Resources

[3]Environmental Engineer, Wis. Department of Natural Resources

Sites which were in violation of the state's locational criteria, including setbacks to navigable surface waters, floodplains, within a wetland or other critical habitat received orders to close. In addition, a change in operations from open, burning dumps to sanitary landfills was enforced at facilities throughout the state. The remaining landfills which posed the greatest threat to the environment due to hydrogeologic setting or poor operations were identified and groundwater and surface water monitoring was required. By the mid to late 1970's groundwater monitoring data from numerous facilities provided documentation that unlined landfills were causing significant groundwater quality impacts and as a result many were required to close. These combined efforts led to the closure of a majority of the 2000 sites which previously existed.

Prior to 1984, Wisconsin rules specified that solid waste disposal facilities be located and operated in a manner that would not cause a "detrimental effect" on groundwater quality. The rules defined detrimental effect as "having a significantly damaging impact on groundwater quality for any present or future consumptive or nonconsumptive uses". Groundwater in Wisconsin is a very important resource since approximately 75% of the state's population relies on groundwater for drinking supplies. Recognizing the importance of this resource, a conservative approach was taken in interpreting the no detrimental effect standard.

In 1984, the Wisconsin legislature, in cooperation with the WDNR and others, passed a comprehensive groundwater law which created specific groundwater quality standards. The old standard of "no detrimental effect" was subject to varying interpretations and was often difficult to defend in judicial proceedings. The new standards were developed using a two step approach. Enforcement standards were generally set to be equivalent to federal drinking water levels and cannot be violated beyond a set distance from a facility (150 feet for new sites and 300 feet for existing ones) or from a facility property boundary, whichever is less. A more stringent trigger level was also established (the Preventive Action Limit) which ranges from 10% to 50% of the enforcement standard depending on the contaminant. This level is applicable at any location where groundwater is monitored, including directly beneath a facility, and requires the owner to "take whatever actions are necessary" to avoid exceeding enforcement standards. Since October 1, 1985, all new landfills must be designed to meet the Preventive Action Limits. A detailed discussion of how Wisconsin's groundwater law affects landfills is provided by Schuff (1986).

Based primarily on groundwater legislation and landfill design and performance experience gained during the 1980's, a new set of solid waste rules were established in 1988 which contained specific design standards, clay liner specifications, construction documentation requirements, as well as expanded performance monitoring requirements.

To implement the regulations, the WDNR now has a plan review staff of 13 engineers and 14 hydrogeologists and a field staff of 24 solid waste investigators/inspectors. Funding for the program and staff is generally provided through plan review and license fees with some supplemental general funds. The necessary regulatory authority to require detailed plan submittals and the routine evaluation of performance data by trained and experienced technical staff have resulted in stringent design and monitoring standards for solid waste landfills in Wisconsin.

Approximately 85 landfills with full containment designs have been approved since 1976. Of this total, nearly 80 are fully operational and a majority of the remainder should be under construction during 1990. These facilities accept over 90% of the municipal and industrial solid waste disposed of in the state and have replaced many of the older landfills that were required to close. A majority of the remaining solid wastes are disposed of in facilities with some degree of engineering controls and groundwater monitoring. As of July 1990, 95% of the solid waste disposed of in Wisconsin is placed in landfills where groundwater is monitored.

The comprehensive monitoring required for the numerous clay-lined landfills in Wisconsin has resulted in a tremendous amount of performance data being generated. The WDNR has routinely summarized and evaluated these data to determine whether the clay liner design concept remains viable. Detailed performance data from 4 operational landfills in Wisconsin showed that thick, well-constructed clay liners can provide a high level of groundwater protection at facilities accepting municipal and many industrial solid wastes (Gordon et al., 1984). More recently, Gordon et al. (1989) calculated hydraulic conductivities of clay liners at 3 full scale landfills based on outflow volumes collected from basin lysimeters located directly below the clay liners. In that study, it was concluded that a field hydraulic conductivity of 1×10^{-7} cm/sec or less can be achieved with thick, well-constructed clay liners. Last year, Huebner et al. (1989) evaluated the quality of liquid collected from a collection basin lysimeter located beneath a clay-lined landfill constructed in 1982. This facility was chosen because it had the most data of any landfill with a basin lysimeter. The evaluation concluded that contaminants had not migrated through the liner during the 7 years the facility has been in operation.

Data from Wisconsin facilities provide field verification that thick clay liners in conjunction with an efficient leachate collection system can provide a high level of groundwater protection. It remains necessary for the WDNR to continue to routinely evaluate landfill performance and to require the necessary design changes to ensure those facilities continue to meet established groundwater protection standards. This paper provides a detailed discussion of Wisconsin's current regulatory approach for landfill design to meet the performance based groundwater standards.

REGULATORY REQUIREMENTS FOR LANDFILL DESIGN

Wisconsin's current solid waste management regulations specify that clay used for liner material meet the following requirements: a minimum of 50% P-200 (silt and clay-size material), a liquid limit of 30 or greater, a plasticity index of 15 or greater, a clay size fraction (0.005 mm) of 25% or greater and a hydraulic conductivity of 1×10^{-7} cm/sec or less. The clay must be compacted to a minimum of 90% modified Proctor or 95% standard Proctor density.

Clay specifications were selected to result in a liner that is primarily silt and clay-size material with good workability and high attenuation characteristics. The specified Atterberg limits favor the use of inorganic soils with lower frost and erosion susceptibility. Because high shrink-swell soils are not common in Wisconsin, no upper limit has been placed on these parameters. Additional justification for codification of these requirements is given by Gordon et al. (1984).

A liner thickness of 1.5 m (5 feet) is required for several reasons. Early experience with clay liner construction in Wisconsin revealed several instances where thinner liners were not adequately constructed leaving little or no barrier at the base of some sites. A 1.5 m (5 foot) liner thickness was selected to avoid this problem by providing a substantial factor of safety to compensate for such construction errors and the difficulty of grading the large aerial extent of a liner to close tolerances. Wisconsin rules also require that the thickness of an individual lift of a liner cannot exceed 15.24 cm (6 inches). This requires that a liner be constructed in numerous lifts and thus minimizes the chance of a truckload of marginal material jeopardizing the overall performance of the liner. The effects of climate and poor operations were also taken into account when the 1.5 m (5 foot) thickness was chosen. Research by Day and Daniel (1985) provides further documentation that thin clay liners can have field hydraulic conductivities much greater than the 1×10^{-7} cm/sec typically specified.

Wisconsin rules require that liner efficiency modeling be performed to optimize leachate collection efficiency as well as to help predict the amount of leachate that will have to be treated. The most frequently used model was initially developed by Wong (1977). The Wong model assumes a uniform liner exists and does not account for any construction variability. The Wong model was modified and further evaluated by Kmet et al. (1981). This analysis showed that even if a uniformly constructed barrier is assumed, a liner thickness of less than 0.6 m (2 feet) would result in a sharp increase in the amount of leakage through the liner. There is considerable reduction in leakage as liner thickness is increased until a point of diminishing return is reached between 1.2 to 1.8 m (4 to 6 feet). This analysis provided WDNR with additional justification for requiring thick clay liners.

A granular drainage blanket must be placed directly above the clay liner for two major reasons. First, the layer provides a uniform, continuous and highly permeable connection between the waste and the leachate collection system. Second, it provides some protection for a liner from desiccation, erosion, freeze-thaw and poor operations. Wisconsin rules require that a drainage blanket be at least 0.3 m (1 foot) thick and consist of either a coarse sand or gravel. The material used for the drainage blanket must have less than 5% P-200 content and a hydraulic conductivity of 1 X 10^{-3} cm/sec or greater.

Several of the early clay liner designs proposed to use refuse in place of soil for drainage blanket purposes. The proposals were initially rejected by the WDNR because this approach would not provide any protection of a liner from desiccation, erosion and freeze-thaw between the time that construction was completed and the time a facility was authorized to begin accepting waste. Further support for this decision is provided by the modeling done by Kmet et al. (1981) which showed that liner leakage is most sensitive to the ratio of hydraulic conductivity of the liner to hydraulic conductivity of the drainage blanket. Data on hydraulic conductivity of municipal refuse confirms that it is typically one order-of-magnitude less permeable than the 1 X 10^{-3} cm/sec requirement for granular drainage blankets (Hughes, 1971; Fungaroli and Steiner, 1979; and WDNR files) and could cause problems with leachate collection efficiency.

A substantial portion of Wisconsin's rules covering landfill design are centered on the leachate collection system. The rules require that all leachate collection systems be designed to limit the post-closure leachate head on a liner to 0.3 m (1 foot) or less. When reviewing proposed designs, estimated leachate head levels are calculated using the liner efficiency modeling described earlier, as well as the Environmental Protection Agency (EPA) water balance method developed by Fenn et al. (1975). Kmet (1982) evaluated the EPA water balance and concluded it was a good tool for predicting leachate generation volumes. His analysis also indicated that input values for runoff coefficients and soil moisture storage have the largest impact on the final results. Many early water balance calculations used values for these parameters which resulted in predictions that little or no leachate would be generated especially after facility closure. Performance monitoring data from existing landfills were used by the WDNR to narrow the range of appropriate values for these parameters and thus result in more realistic calculations of leachate generation. In addition, Scharch (1984) computerized the EPA water balance method to allow for rapid evaluation of the sensitivity of input parameters and to assist in overall design of the leachate collection system.

While modeling efforts are relied upon to assist in the design of leachate collection systems, Wisconsin rules specify the following minimum design requirements: liner base slopes must be at least 2%; collection pipes must be Schedule 80 PVC or equivalent with a minimum diameter of 15.24 cm (6 inches) and must be placed in a trench excavated into the liner; cleanout access points must be designed on both ends of the collection lines and the whole system must be designed to provide gravity drainage of leachate to a storage tank or directly to a sanitary sewer. While some of these design criteria result from application of standard engineering principles, the rest have evolved and continue to be improved upon based on performance monitoring of existing containment facilities. The development of leachate collection system design in Wisconsin is described in detail by Kmet et al. (1988).

The last major element in the design of an engineered landfill is the final cover system. Wisconsin administrative rules written in 1980 required a final cover consisting of 0.6 m (2 feet) of fine-grained soils overlain by 0.15 m (0.5 feet) of topsoil. In a detailed evaluation of these requirements, Grefe et al. (1987) concluded that a thicker final cover is necessary to assure long-term protection of the compacted clay layer and to sustain a healthy vegetative cover on landfills. In addition, previous regulations allowed final slopes of as little as 2% and as steep as 33%. Slopes which were built at 2% ultimately became flat or concave due to settlement of the waste mass while slopes constructed at 33% were difficult to vegetate and typically experienced local slope failure and excessive erosion. These findings led to rule revisions which include requirements for a 0.45 to 0.75 m (1.5 to 2.5 foot) uncompacted cover soil layer to be placed over the clay capping layer, minimum slopes of 5% and maximum slopes of 25%.

CONSTRUCTION DOCUMENTATION

To ensure that a well thought out engineering design is properly constructed in the field, it is necessary to provide engineering oversight during construction of the facility. As with early landfill design specifications, requirements for construction documentation were also developed on a case-by-case basis as part of the approval of final engineering plans for a facility. While these requirements varied somewhat initially, they became quite consistent and relatively well accepted by the mid 1980's. Current rules require that a registered professional engineer or an engineering technician under the direct supervision of a registered professional engineer be present during all critical construction activities. This oversight consists of providing detailed surveying to document liner and drainage blanket thickness, proper liner slope, leachate collection system layout and all other associated piping and structures.

The liner and drainage blanket must be tested on a regular basis to confirm that the material being delivered to the site meets or exceeds the project specifications and that it is being placed in accordance with state regulations. Table 1 contains a summary of the specific testing requirements.

TABLE 1. Soil Testing Requirements for Clay Liner and Drainage Blanket Construction

Sampling method (1)	Testing frequency (2)	Properties to be determined (3)
	(a) Clay Liner	
Nuclear density meter	5 tests/acre/lift	Field density
Nuclear density meter	5 tests/acre/lift	Field moisture content
Grab	Every 5000 yd^3	Moisture-density curve, grain size and Atterberg limits
Undisturbed	1/acre	Grain size, Atterberg limits, moisture content, dry density and hydraulic conductivity[1,2]
	(b) Drainage Blanket	
Grab	Every 1000 yd^3	Grain size
Lab	Every 2500 yd^3	Moisture content, dry density and hydraulic conductivity

[1]Testing only required every third sample.
[2]WDNR may require that a portion of the hydraulic conductivity tests be performed using leachate.

Once construction has been completed, a report along with a set of as-built engineering plans must be prepared which documents all aspects of facility construction. The report and plans must be submitted to the WDNR for review and disposal of solid waste cannot begin until the report is approved and the facility has been inspected.

Wisconsin rules require extensive density and moisture content testing during placement of the clay. This testing along with engineering oversight by both an engineering consultant firm as well as WDNR staff are heavily relied upon to ensure that a uniform and dense low hydraulic conductivity clay liner is constructed in the field. Monitoring data have shown that a properly constructed thick clay liner which meets or exceeds the grain size, Atterberg limits and density requirements can achieve a field hydraulic conductivity of 1×10^{-7} cm/sec or less (Gordon et al., 1989 and WDNR files).

PERFORMANCE MONITORING AND RESULTS

After a facility has been constructed and disposal operations begin, a comprehensive operational and long-term monitoring program is required to document whether a site is performing as designed. One simple method of documenting performance is to measure and record the volume of leachate extracted from a site on a regular basis. This is accomplished by recording volumes each time a leachate holding tank or lift station is pumped or by use of a flow meter if the facility has a direct connection to the sanitary sewer.

Figure 1 shows monthly volumes of leachate collected at a clay-lined landfill which began operations in December 1978 and was closed in the fall of 1989. As shown in this figure, leachate was generated at the site within the first year of initiating operation. This is typical for clay-lined municipal solid waste landfills located in Wisconsin where annual precipitation averages slightly more than 76.2 cm (30 inches) per year. Table 2 contains more detailed data regarding leachate generation at the facility. The Table shows that the amount of leachate collected at the site typically ranged between 6.35 and 8.89 cm (2.5 and 3.5 inches) per year. Some of the decrease in the amount of leachate collected at the site in the last two years is the result of below normal rainfall. In the area of this particular landfill, rainfall was 12.70 cm (5 inches) below normal in 1988 and 15.24 cm (6 inches) below normal in 1989.

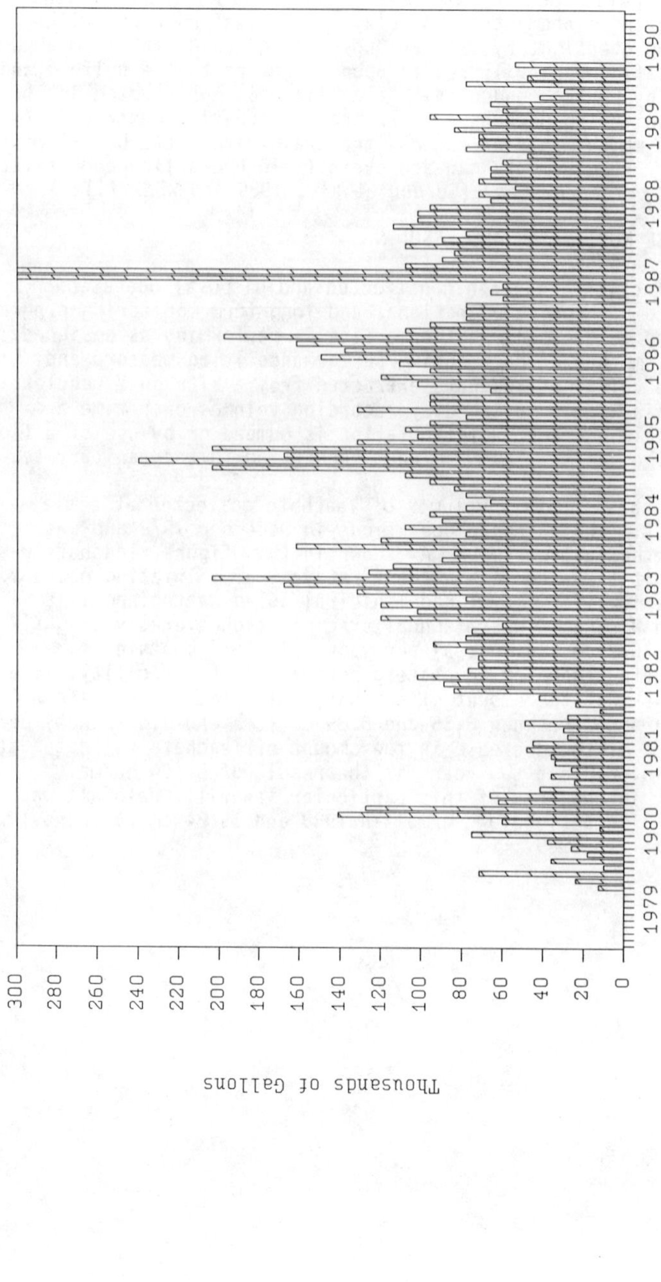

Fig. 1. Monthly Leachate Generation Volumes at the Eau Claire County Landfill

TABLE 2. Summary of Leachate Generation at the Eau Claire County Landfill

Year (1)	Leachate Volume Collected (gallons) (2)	Average Area Contributing to Leachate Generation (Acres) (3)	Leachate Collected (inches/year) (4)
1979	108,000	3.7	1.08
1980	671,500	7.0	3.53
1981	443,000	7.0	2.33
1982	824,000	10.3	2.95
1983	1,389,000	13.7	3.73
1984	1,140,000	13.7	3.06
1985	1,668,000	18.0	3.41
1986	1,443,000	18.0	2.95
1987	2,088,020	24.0	3.20
1988	978,00	24.0	1.50
1989	774,000	24.0	1.19
1990[1]	330,000	24.0	1.01[2]

[1]Includes data from January through June
[2]Projected annual amount

Wisconsin rules require that leachate head levels within a landfill be measured. In order to measure leachate head levels, wells are either installed as part of the initial construction of the facility and extended upward as filling progresses or are drilled after an area reaches final grade. In either case, the wells must be installed directly above the clay liner but without penetrating it. Leachate collection systems must be designed to maintain a post-closure leachate head level of less that 0.3 m (1 foot) on the base of the clay liner. A majority of the available data are from measurements at landfills which are still in operation. This information indicates that leachate head levels at operating facilities range between 0.6 to 0.9 m (2 to 3 feet). Not enough data currently exists to reach a conclusion regarding actual head levels at closed facilities.

Excessive head levels may indicate that leachate is not readily reaching the collection system. This could be due to clogging of the drainage blanket or because of physical damage to one or more of the collection lines. Problems with leachate collection systems are discussed in detail by Bass (1984) and Kmet et al. (1988). After closure, increased leachate collection rates or excessive leachate head levels within a landfill may indicate that the final cover system is not performing properly or that the leachate is not being extracted on a routine basis.

Leachate head wells must be protected against damage during filling and from closure activities. When installed after an area reaches final grade, care must be taken to ensure that the clay liner is not penetrated during drilling. In Wisconsin, a minimum of 2 leachate head wells must be installed in each major horizontal phase of a facility. A major phase is defined as a horizontal portion of a landfill designed to be constructed at one time.

Samples of leachate are required to be collected on a routine basis and analyzed for both organic and inorganic parameters. Sridharan and Didier (1988) summarized available leachate quality data from 56 of the containment landfills in Wisconsin and found that almost 50% of the median values for chemical constituents identified, exceeded U.S. drinking water standards and Wisconsin groundwater enforcement standards. This provided further confirmation that leachate containment and collection was necessary to protect groundwater quality.

To provide an early indication of the quantity and quality of liquid seeping through a clay liner, all major horizontal clay-lined phases of facilities located above the groundwater table must be designed with a collection basin lysimeter to monitor the unsaturated zone. Collection basin lysimeters afford a practical means to quantify seepage through a discrete portion of a liner system and provide data to calculate transit times of contaminants passing through a clay liner (Daniel, 1987; Grube et al., 1987; and Reades et al., 1987). A collection lysimeter is a basin constructed with a geomembrane and geotextiles are usually installed above and below the geomembrane to protect it from damage or puncture. A perforated pipe is installed at the bottom of the basin and a solid transfer pipe exits the lysimeter through a prefabricated boot. Liquid migrating through the portion of a clay liner above the lysimeter accumulates in the basin and flows by gravity to a collection point where the volume can be measured and samples can be taken for analysis.

A generalized cross-section of a collection basin lysimeter is shown in Figure 2. These devices are required to be leak tested over a 24 to 48-hour period to ensure they are water tight. Additional information on their design and construction is provided by Kmet and Lindorff (1983). Collection basin lysimeters are required to be monitored quarterly. The volumes extracted must be recorded and samples of the liquid must be analyzed for the same organic and inorganic parameters required for monitoring leachate quality.

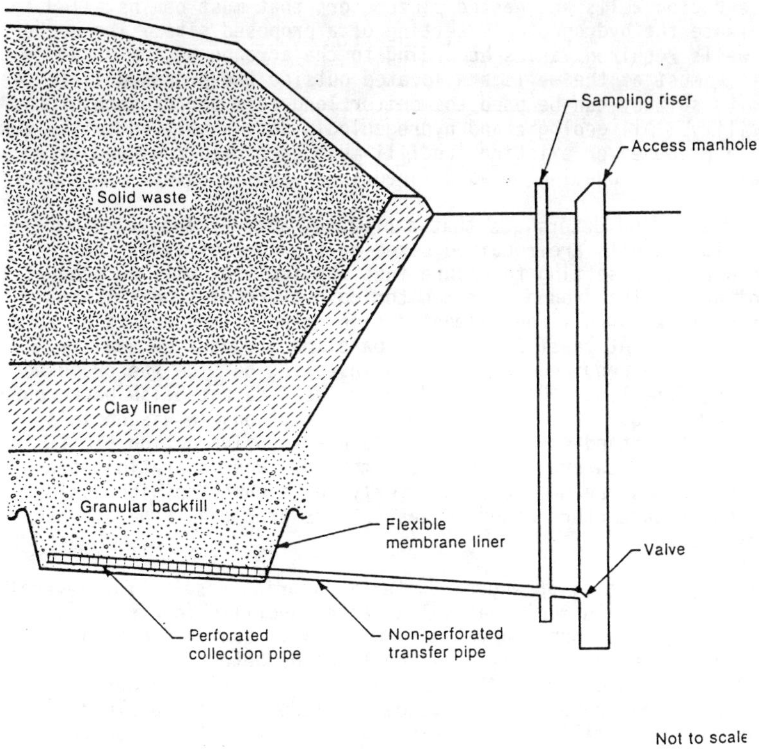

Figure 2. Longitudinal Cross-section of a Collection Basin Lysimeter

Wisconsin rules specify the minimum number of water table observation wells and nested piezometers that must be installed to evaluate the hydrogeologic setting of a proposed site. The number of wells required varies according to the acreage of the proposed site. Most of the wells are located outside the proposed fill limits so they can be used to monitor long-term performance of the facility. All geologic and hydrogeologic data submitted to the WDNR for a proposed or existing landfill must have been reviewed by a qualified hydrogeologist as defined in Wisconsin rules.

Once the WDNR determines that a proposed site is feasible, additional wells are required based on the complexity of the hydrogeologic setting to ensure that the facility will be adequately monitored. The groundwater monitoring network at landfills in Wisconsin with full containment designs generally consists of 20 or more wells. At least 8 rounds of baseline groundwater quality sampling are required on all wells installed outside the proposed limits of filling. Resulting data are used to establish background groundwater quality, the need for any exemptions to Wisconsin's groundwater standards and to calculate Preventive Action Limits for inorganic indicator monitoring parameters. When operations begin, all wells must be monitored quarterly for a number of inorganic indicator parameters based on waste types proposed to be accepted at the site.

In reviewing available groundwater monitoring results for several of the clay-lined municipal solid waste landfills in Wisconsin, Friedman (1988) found only 1 monitoring well containing volatile organic compounds (VOCs) above the level of detection. This occurred at an older landfill not built to today's standards in Wisconsin and therefore, the study concluded that properly designed clay-lined landfills could achieve a high level of groundwater protection.

All new landfills in Wisconsin must be designed to provide adequate drainage and sedimentation controls to ensure that any nearby surface waters or wetlands are not impacted. Any surface water runoff which comes in contact with solid waste is required to be collected and treated as leachate. Facilities which discharge runoff from a sedimentation basin to a surface water or wetland must monitor that discharge for field conductivity, field pH, biological oxygen demand (BOD), and total suspended solids on a weekly basis whenever water is actively discharging from a basin.

A series of gas probes must be installed outside the limits of filling on all sides of a facility in deposits of granular soil and any other soil formations where gas migration may occur. Inspections for vegetative stress and measurement of temperature, ground condition, barometric pressure and information as to whether barometric pressure is rising or falling is required to be included in the gas monitoring program.

Gas produced within a municipal solid waste landfill must be collected and combusted to control the release of any hazardous air contaminants. If the landfill is located in a populated area and it is determined that excessive particulate emissions will be caused by landfill construction and operation, a program to provide periodic air quality monitoring is required.

CONCLUSION

Wisconsin has a strong and comprehensive regulatory program for solid waste management. The WDNR has required the use of clay liners as a means of leachate containment and groundwater protection at solid waste landfills for over 15 years. Within this timeframe, considerable practical design, construction and operational experience has been gained. Routine evaluation of performance monitoring data has also resulted in further refinements in landfill design criteria and monitoring requirements.

Based on Wisconsin's experience with clay liner design and evaluation of performance monitoring data from numerous facilities, it is concluded that properly designed and constructed clay liners along with an efficient leachate collection system can provide a high level of groundwater protection at solid waste disposal facilities. However, the technology remains developmental and may require further modifications in the future as additional performance data become available.

REFERENCES

Bass, J.M., Ehrenfeld, J.R., and Valentine, J.N. (1984). "Potential clogging of landfill drainage systems. U.S. EPA. NTIS Order No. PB-84-110 550.

Daniel, D.E. (1987). "Hydraulic conductivity tests for clay liners." Proceedings of the Ninth Annual Symposium on Geo-Aspects of Waste Management. Fort Collins, CO., pp. 15-34.

Day, S.R., and Daniel, D.E. (1985). "Hydraulic conductivity of two prototype clay liners." Journal of Geotechnical Engineering, ASCE, 111(8), pp. 957-970.

Fenn, D.G., Hanley, K.J., and DeGeare, T.V. (1975). "Use of the water balance method for predicting leachate generation from solid waste disposal sites." EPA/530/SW-168. U.S. EPA, Cincinnati, OH. 40 pp.

Friedman, M.A. (1988). "Volatile organic compounds in groundwater and leachate at Wisconsin landfills." Wisconsin Department of Natural Resources PUBL-WR-192, 79 pp.

Fungaroli, A.A., and Steiner, R.L. (1979). "Investigation of sanitary landfill behavior - volume 1." Final Report EPA-600/2-79-053a.

Gordon, M.E., Huebner, P.M., and Kmet, P. (1984). "An evaluation of the performance of four clay-lined landfills in Wisconsin." Proceedings of the Seventh Annual Madison Conference of Applied research and Practice on Municipal and Industrial Waste, pp. 399-460.

Gordon, M.E., Huebner, P.M., and Miazga, T.J. (1989). "Hydraulic conductivity of three landfill clay liners." Journal of Geotechnical Engineering, 115(8), pp. 1148-1160.

Grefe, R.P., Huebner, P.M., and Gordon, M.E. (1987. "Multilayered cover design and application to Wisconsin landfills." Proceedings of the Tenth Annual Madison Waste Conference of Applied Research and Practice on Municipal and Industrial Waste, pp. 1-32.

Grube, W.E., Roulier, M.H., and Hermann, J.G. (1988). "Implications of current soil liner permeability research results." Proceedings of the Thirteenth Annual Research Symposium on Land Disposal, Remedial Action, Incineration and Treatment of Hazardous Waste. (EPA/600/9-87/015), U.S. EPA Hazardous Waste Engineering research Laboratory, Cincinnati, OH. pp. 3-25.

Huebner, P.M., Gordon, M.E., and Miazga, T.J. (1989). "An evaluation of the quality of seepage from a clay-lined landfill." Proceedings of the Twelfth Annual Madison Waste Conference of Applied Research and Practice on Municipal and Industrial Waste, pp. 78-97.

Hughes, G.M. (1971). "Hydrogeology of solid waste disposal sites in northeastern Illinois." Final Report. U.S. EPA SW-12d.

Kmet, P. Quinn, K.J., and Slavik, C. (1981). "Analysis of design parameters affecting the collection efficiency of clay-lined landfills." Proceedings of the Fourth Annual Madison Waste Conference of Applied Research and Practice on Municipal and Industrial Waste. pp. 204-227.

Kmet, P. (1982), "EPA's 1975 Water Balance Method -- It's use and limitations." Wisconsin Department of Natural Resources Internal Staff Memorandum. 48 pp.

Kmet, P., and Lindorff, D.E. (1983). "Use of collection lysimeters in monitoring sanitary landfill performance." Proceedings of the Characterization and Monitoring of the Vadose (unsaturated) Zone. NWWA. pp. 554-579.

Reades, D.W., Pohland, R.J., Kelly, G., and King, S. (1987). Discussion, Journal of Geotechnical Engineering, vol. 113, no.7, pp. 809-813.

Scharch, J.F. (1985). "Water balance analysis program for the IBM PC micro computer." Bureau of Solid Waste Management, Wisconsin Department of Natural Resources. 68 pp.

Schuff, R.G. (1986). "Wisconsin's groundwater law - how it affects regulation of solid waste landfills." Proceedings of the Landfill Management Program, Waste Tech '86. Chicago, Il., pp. 15-25.

Sridharan, L., and Didier, P. (1988). "Leachate quality from containment landfills in Wisconsin." Proceedings of the Fifth Annual International Solid Wastes Conference. ISWA, vol. 2, pp. 134-138.

Wong, J. (1977). "The design of a system for collecting leachate from a lined landfill site". Journal of Water Resources Research, vol. 13, No. 2.

Wisconsin Department of Natural Resources Files

v:\perm\sw9ascsf.meg

Performance Evaluation of Earthen Liners

Stephen J. Trautwein[1], M. ASCE and Charles E. Williams[2], M. ASCE

Abstract

Past and current methods of evaluating the performance of earthen liners are reviewed. Laboratory testing of small diameter samples in the design phase was the primary means of performance evaluation in the past. Current performance evaluation involves testing in the design, construction, and service phases of a liner project. Emphasis is on large scale in situ testing in order to account for the impact of secondary features on hydraulic conductivity. Several case histories are presented in which the liner contained obvious secondary features. In situ hydraulic conductivity measured using a Sealed Double Ring Infiltrometer was between three and one hundred times that measured in the laboratory using small diameter samples. A fourth case history was presented in which no visible secondary features were detected in the liner. Hydraulic conductivity values determined in situ and in the laboratory agreed well.

Introduction

Engineered earthen liners are a key component of containment systems underlying most modern landfills. Containment systems can consist of single or multiple earthen

---

[1]President, Trautwein Soil Testing Equipment Co., P. O. Box 31429, Houston, TX 77231

[2]Vice President - Environmental Engineering, Sanifill, Inc. Houston, TX

liners. In addition, manufactured liners, geomembranes and geobentonite composites, are often used together with earthen liners. Together or singly, these liners or barriers can eliminate leakage or minimize the rate at which leachate within the waste facility escapes into the surrounding environment.

The effectiveness of an earthen liner as a barrier depends on the materials it is comprised of and how well the liner is constructed. Earthen liners can be constructed from a variety of soil types and mixes of soil types. The material can be placed at a wide range of density and moisture content combinations. Many different types of equipment can be used and conditions during placement can range from extreme hot to extreme cold, wet to dry, and windy to calm.

The wide range of factors listed above make it challenging to choose a combination which results in an effective earthen liner. The potential for error in material control and construction control makes performance evaluation all the more important. Included in this paper are discussions on the approach to performance evaluation, both past and present. Verification testing techniques are presented and evaluated. Several case histories demonstrating the importance of performance evaluation are given. Finally, suggestions for future improvements in performance evaluation are discussed.

Performance Evaluation

Performance evaluation is the process by which a product is tested and examined to ensure that it operates or behaves as it was designed to. Part of that process involves verification testing, i.e., subjecting the product to actual conditions, or if possible, conditions more severe than service conditions. Ideally, the entire product is tested under actual conditions. Performance evaluation then is a straightforward process. The product passes or fails. If it passes, then the product can be put into service with a high degree of confidence that it will perform properly.

In other cases, practical considerations limit verification testing to a small portion of the product. Also, test conditions may be less severe than service conditions. In the absence of definitive verification testing, performance evaluation involves using engineering judgement. In these cases, the product is put into service with a prediction as to how it will perform. The level of confidence in that prediction is directly related to the amount of verification testing and inversely related to the amount of engineering judgement involved in the performance evaluation.

Performance evaluation for earthen liners fall into this latter category. The large size of most liners deem it impractical and too costly to perform full scale testing. Even if full scale testing were possible, subjecting the liner to the environmental, stress, and deformation states, expected to occur over the life of the landfill, would be impractical. As a result, performance evaluation of earthen liners is based largely on engineering judgement combined with limited verification testing.

Performance Evaluation - Past

Influencing Factors

Performance evaluation in the past was influenced by the opinion that little or no leakage would occur through liners constructed from soils with hydraulic conductivity less than 1x10-7 cm/s. This is evidenced by the fact that statements found in soil reports during the 1950's and 1960's included specifications such as "the liner material shall be constructed using an impermeable clay soil". As a consequence, little verification testing was performed. The little that was is described below.

Verification Testing

Nearly all the testing used to evaluate the hydraulic conductivity of earthen liners in the past consisted of laboratory tests on small diameter (25 mm - 100 mm) cylindrical samples. Testing was generally performed in the design phase to determine if a soil could be compacted such that the hydraulic conductivity was less than 1x10-7 cm/s. If it could, it was assumed the hydraulic conductivity of the liner would be the same as that determined in the laboratory.

The majority of the tests were performed on samples recompacted in the laboratory. In relatively few cases, verification testing was performed on "undisturbed" samples obtained using sampling tubes that were either pushed or driven into the liner. Testing was conducted in both rigid-wall and flexible-wall permeameters. In some cases, consolidation equipment was used to back calculate hydraulic conductivity using consolidation theory. Testing procedures varied widely as there were no universally accepted standard procedures to follow.

With no data to suggest that liners leaked, there was no incentive to do much else in the way of verification testing. However, by the late 1970's, it started to become apparent that some earthen liners were leaking at rates much higher than design rates. This led to a closer examination of earthen liners which in turn led to a better understanding of factors that affected performance. These factors are discussed below.

Factors That Affect Performance

Daniel, 1984, examined several existing earthen liners for which leakage rates could be quantified. Field hydraulic conductivity values were back calculated from these observed leakage rates and were found to be 5 to 100,000 times greater than those determined in the laboratory. This study led other investigators to construct test pads (small sections of earthen liners) for which laboratory and field hydraulic conductivity values could be compared (Day et al., 1985, Rogowski 1986, Elsbury et al., 1990). The major findings from these studies were: (1) laboratory tests on small samples could yield hydraulic conductivities much less than field values; and (2) secondary features controlled the rate of leakage through an earthen liner.

Secondary Features

The ideal soil structure for an earthen liner is one that is uniform, where soil grains are mostly in contact with adjacent soil grains, and the void space or micropores between soil grains has been minimized. While some earthen liners may have a soil structure that approaches the ideal case, most contain macropores or secondary features which transmit greater amounts of flow than micropores. For purposes of discussion, secondary features will be categorized as follows: macro and structural.

Macro secondary features include large interconnected pathways which result from a lapse in Construction Quality Assurance/Construction Quality Control (QA/QC). Examples include an unnoticed change in soil type, presence of deleterious materials such as grass, roots, branches, grade stakes etc, or desiccation cracks. Structural secondary features are more subtle and result from a combination of soil conditions and construction procedures. Examples are clod interfaces, lift interfaces, minor cracking due to desiccation between lifts, and shear surfaces resulting from local bearing capacity failures under the weight of the compaction equipment.

While the hydraulic conductivity of the material bounded by secondary features may be quite low, secondary features can conduct enough flow to increase the apparent hydraulic conductivity of the overall liner by several orders of magnitude. Flow can be quite circuitous, with flow patterns being similar to that which occurs in fractured rock.

Secondary features in many liners were the result of using the same soil specifications, procedures, equipment, and quality control used for conventional earthwork projects (e.g. roadway subgrades, and embankments). The key objective in placing soil in these projects was to maximize strength and minimize the compressibility, not minimize the hydraulic conductivity. Conventional earthwork structures can include secondary features and distributed marginal areas without affecting overall performance. Earthen liners on the other hand need to be constructed using more rigorous procedures designed to eliminate or minimize the presence of secondary features.

Construction procedures needed for earthen liners but not necessarily for conventional earthwork projects include: (1) screening of the soil to remove large particles; (2) use of pulverizers to break down soil clods; (3) use of tillers to mix the soil when moisture adjustment is required; (4) use of deep footed rollers to provide kneading compaction and elimination of lift interfaces; and (5) protection of the upper surface of the liner to prevent desiccation

Another important point to note is that the key parameter in QA/QC for conventional earthen projects is density, a good indicator of strength but a poor indicator of hydraulic conductivity. For example, in laboratory studies, it has been shown that kneading compaction resulted in a hydraulic conductivity four times less than that obtained using static compaction, even though the samples had similar densities (Mitchell et al., 1965). For earthen liners, compaction moisture content is more important than density. Moisture content controls the strength of the clods and hence the ability to knead the soil which helps to eliminate or minimize structural secondary features.

Laboratory Testing

Hydraulic conductivity values measured in the laboratory were often less than those measured in the field. The main reason for this was that laboratory samples were not representative of the in situ soil. While the material may have been the same, the difference was the structure. Laboratory samples were too small to account for secondary features present in the liner. Flow in the field was through secondary features; flow in laboratory samples was through micropores. Even if laboratory samples contained secondary features, the samples were typically too small in diameter for the flow channel to be continuous through the sample.

Other factors that could result in hydraulic conductivity values measured in the laboratory being less than those measured in the field include:

(1) Compressive stresses imparted to the soil during the sampling process can reduce the size of macro pores
(2) Sample selection - technicians often discard samples with obvious defects and opt to test more intact samples.
(3) Smearing over macro pores at each end of the sample during the trimming process.
(4) A reduction in the size of macro pores due to the use of confining pressures in the laboratory that are much higher than those in the field.

It should also be noted that without a commonly accepted standard procedure for performing laboratory hydraulic conductivity tests, a wide variety of equipment and procedures were used. Regulatory agencies seldom required more than reporting of a value for hydraulic conductivity. No specifications as to equipment type, gradient, or confining pressure were given. Hence, a test performed in a flexible-wall cell with backpressure, low gradient, and low confining pressures carried the same weight as a test with no backpressure, high gradient, and high confining pressure.

Performance Evaluation - Current

Influencing Factors

The studies mentioned in the previous section had a major impact on the thinking of the regulatory community. As a result, improvements were required in the performance evaluation process. These improvements included additional verification testing to take place not only in the design phase, but during the construction and service phases of the project as well. Also, emphasis was placed on in situ testing because of concern about the reliability of laboratory hydraulic conductivity testing.

The most commonly used verification testing techniques are discussed below.

Verification Testing

Laboratory testing. As pointed out previously, the results of laboratory hydraulic conductivity tests on small samples can be grossly misleading when used to estimate the in situ hydraulic conductivity. In spite of this, such testing continues to be part of standard practice and may persist for a time as convention is difficult to change. The authors strongly recommend that if laboratory testing of small samples is to be used, other types of verification testing be performed as well.

One alternative is to perform laboratory hydraulic conductivity tests on larger samples. The criticism of laboratory hydraulic conductivity tests is due to sample size, not the test method. There are important advantages that make laboratory hydraulic conductivity testing attractive for verification testing. These advantages include shorter testing times and the ability to control boundary conditions, stress levels, and degree of saturation. Also, samples can be taken without delaying construction. The question is, how large of a sample is needed for test results to reflect in situ hydraulic conductivity? The answer is, it depends on the amount and type of secondary features present in the liner. If macro secondary features are present, the sample size needed would be too large to test in the laboratory. If only structural secondary features are present, then a sample size practical for handling in the laboratory may yield hydraulic conductivity values which reflect in situ values.

The authors constructed a flexible-wall permeameter that can accommodate samples with diameters of up to 300 mm. This has been used to test samples from several liner projects. Results from two of those projects are discussed in a later section.

Another area of concern about laboratory testing is the lack of a widely adopted standard test procedure. This situation should improve as the American Society for Testing and Materials (ASTM) has adopted a standard test method for determining the saturated hydraulic conductivity of a soil specimen using a flexible-wall permeameter (ASTM D5051).

Field testing. Starting in the mid 1980's, regulatory agencies began to emphasize in situ testing as a means for verification testing. The main attraction of the in situ testing approach was the ability to test larger volumes of soil with minimum sample disturbance. Testing larger volumes of soil increases the chance of detecting the impact of secondary features on hydraulic conductivity.

A number of devices have been proposed for in situ testing. Those most commonly used can be listed in the following four categories: porous probes, boreholes, infiltrometers, and underdrains. Comments are presented about the usefulness of these devices in the following paragraphs. For more detailed review, the reader is referred elsewhere (Daniel 1989).

Porous probe devices offer moderate cost, convenience, and rapid testing (hours to days). However, these advantages are at the expense of sample size. The volume of soil permeated is typically less than that used in a conventional laboratory test. The most commonly used porous probe device is the BAT probe. (Torstensson 1984).

Borehole tests offer low cost, testing times in the range of several days to weeks, and permeate volumes of soil 30 to 50 times that used in a conventional laboratory test. The tested soil volume is approximately the size of a basketball. The most commonly used borehole devices are the Boutwell permeameter (Soil Testing Engineers 1983), and the Guelph permeameter (Reynolds and Elrick 1985; Soil Moisture Equipment Co. 1987).

Infiltration tests are more expensive to perform and have testing times in the range of several weeks to months. They offer the advantage of being able to permeate large volumes of soil. Infiltrometers require special modifications to ensure accurate measurement of the low flow rates associated with earthen liners. Also, what to use for suction at the wetting front in the determination of hydraulic conductivity is an unresolved issue. The most commonly used infiltrometers are the Sealed Double Ring infiltrometer (SDRI) (Daniel and Trautwein 1986, Trautwein 1989) and the air-entry permeameter (Bouwer 1966).

Underdrains are moderate in cost, have testing times in the range of several months to a year, and can test very large volumes of soil. Underdrains or lysimeter pans are generally constructed in the field, and vary in size and shape. (Gordon et al., 1989, Elsbury 1990).

Testing Phases

In the past, verification testing was commonly performed in the design stage of a liner project. Current performance evaluation involves verification testing not only in the design phase, but also during the construction and service phases. Testing for each phase of the project is discussed below.

Preconstruction Phase. The benefit of preconstruction performance evaluation is that important information is gained in the design phase of the project. Design specifications can be changed in order to achieve the optimum results in the field. Also, regulators can be assured that the proposed design is adequate before permits are granted for liner construction.

This preconstruction testing is generally performed on test pads or test strips (U. S. EPA 1986). Test pads allow for experimentation with different moisture contents, equipment types, and procedures in order to produce a liner with minimal secondary features. They also afford the opportunity to devise and check the QA/QC plan. Field personnel can become familiar with how soil compacts and how to detect problems. The test pad approach also educates contractors in procedures of which they are unaccustomed. This is particularly useful if the contractor's primary experience is in building structural fills.

Construction Phase. Verification testing during the construction phase is limited mainly to laboratory testing. More reliable testing is problematic as time delays are incompatible with construction schedules.

The lack of a quick and reliable method for this critical phase makes it extremely important that a proven QA/QC program be in place. The importance of having experienced, well trained field personnel overseeing liner construction cannot be over emphasized.

Service Phase. This phase is often ignored in the evaluation program. Verification during the service phase can be performed using monitor wells and underdrains. Monitor wells are useful for detecting leakage but not for determining the leakage source or rate. Underdrains are the most effective means of detecting leaks early and quantifying leakage rates.

Recommendations.

Experience on over 50 projects has led the authors to conclude that sample size is the most critical factor in judging the merit of any particular verification test. Testing devices unable to account for the affect of secondary features on hydraulic conductivity, are of little value as the purpose of verification testing is to detect defects.

Test methods which allow for the direct determination of leakage rates are also preferred. Earthen liners are built to control leakage, so why not evaluate performance by leakage rate? The advantage is that the uncertainties associated with determining hydraulic conductivity are eliminated.

For all the in situ methods mentioned above, hydraulic conductivity is determined indirectly. Flow measurements are made and then using simplifying assumptions about boundary conditions, gradient, and degree of saturation, hydraulic conductivity is calculated. There is always a degree of uncertainty in hydraulic conductivity which depends on the simplifying assumptions made.

Leakage rates give the designer information on what is being achieved and regulators a better idea for what is being enforced. Further understanding of liner performance may result if regulations are written in terms of leakage rates as well as hydraulic conductivity.

The technique that best fits the criteria listed above, large sample size and direct measurement of leakage rates, is underdrains. There are no significant experimental errors involved (with the exception of leakage through the bottom of the underdrain) and no theoretical assumptions violated. The main problem with underdrains is the time needed to reach maximum leakage rate. Since exfiltration is measured, leakage rates will be minimal early in the test and then increase with time (Fig. 1). To ensure that maximum leakage rate is determined the test must continue until exfiltration becomes constant. Due to long testing times, underdrains are not commonly used for the preconstruction or construction phases.

If underdrains can not be used, then infiltrometers are preferred, the SDRI in particular (Fig. 2) because of its large size. Infiltrometers are recommended as this test simulates conditions for which the liner is being designed, i.e. ponding of liquid on the surface. In addition, when lateral flow is minimal and there is no soil swell, a direct measurement of the leakage rate can be made. (NOTE: In cases where soil swell occurs, leakage rates will be overestimated because some of the flow being measured is held by the soil and does not flow through the liner). Testing time can be shorter than that required for underdrains, particularly if it is only necessary to demonstrate that the hydraulic conductivity or leakage rate is less than a specified value. Infiltration rates generally decrease with time, thus if the hydraulic conductivity is less than that required and infiltration is decreasing, the test can be terminated (Fig. 1).

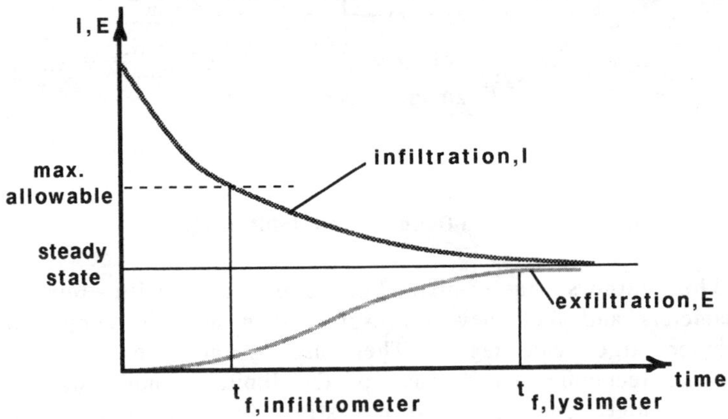

Fig. 1. Infiltration and Exfiltration Rate versus Time

One gap in the current approach to performance evaluation is that little or no verification testing is performed during construction. Infiltrometers can be used beforehand, and underdrains afterwards but no technique has been accepted for testing during construction. The current thinking is to invest a good deal of effort in the preconstruction evaluation, and then rely on QA/QC during construction to provide the same results.

The authors suggest that laboratory hydraulic conductivity tests on large diameter samples (300 mm) may fill this gap. The appeal of this technique is that testing can be performed without delaying construction. What still needs to be proven is whether or not sample diameters of 300 mm are large enough to account for the presence of secondary features.

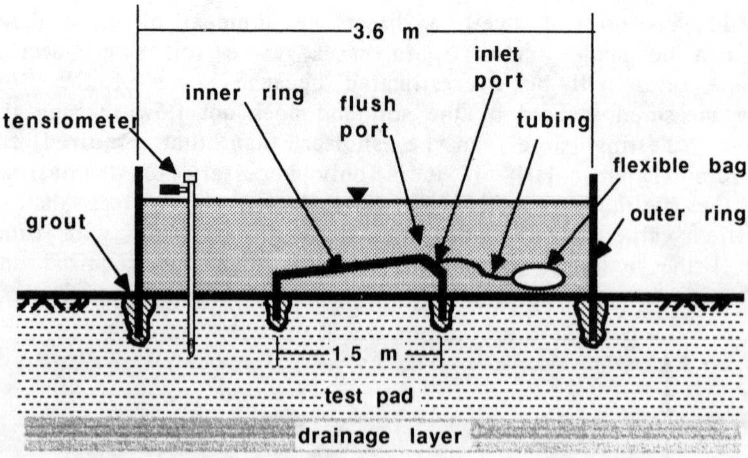

Figure 2. Sealed-Double Ring Infiltrometer

The authors have also had experience with borehole permeameters and feel they are useful when used in conjunction with other large scale tests. Their use as the sole verification test is not recommended because of the limited sample size. If time and budget constraints do not allow for large scale testing with infiltrometers, a combination of large diameter laboratory tests and borehole tests would be recommended. A comprehensive testing program involving a large number of tests and more than one test method will have a better chance of detecting secondary features. However, there is no substitute for high quality testing procedures.

Case Histories

The following case histories have been selected as examples of the importance of field testing and sample size in a performance evaluation study. They also demonstrate the benefits of using a proper QA/QC plan.

Refining Construction Specifications

A test pad was constructed using a soil with properties shown in Fig. 3. The soil was compacted in four 150 mm lifts. Quality control during construction was average. Construction specifications were typical of those used for many landfills, i.e., density equal to or greater than 95% maximum density from a standard proctor test and moisture content at or above optimum. The soil was processed to reduce clod size and particles larger than 25 mm were screened off.

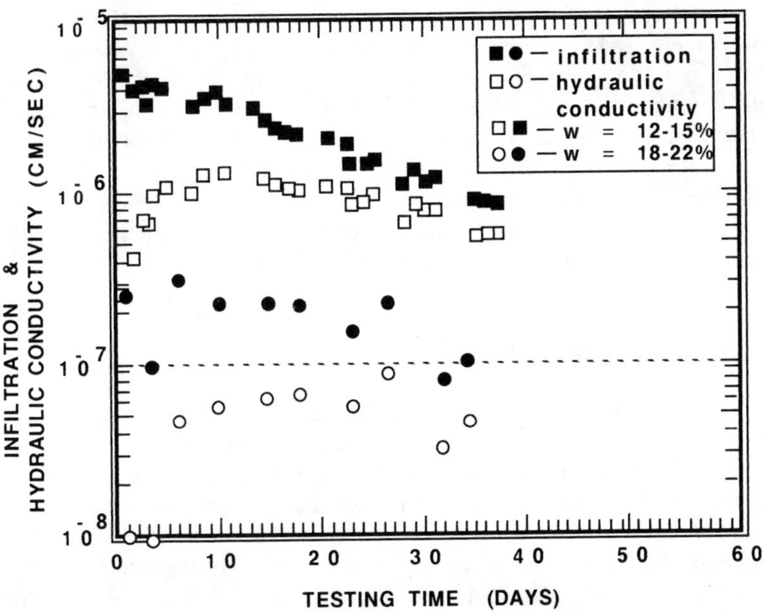

Fig. 3. Sealed Double Ring Infiltrometer Test Results
Soil Properties: PI = 20, W(opt) = 12.5%

Laboratory hydraulic conductivity tests on 75 mm diameter sample yielded values in the $2x10^{-8}$ cm/s range. An infiltration test was performed on the test pad using an SDRI. Tensiometers were used to determine the depth of the wetting front during the course of the test. The resulting infiltration rate and hydraulic conductivity determined from the test is plotted versus time and is shown in Fig. 3. The average hydraulic conductivity after 40 days of testing was $6x10^{-7}$ cm/s, 30 times that determined from laboratory tests.

A trench was excavated through the test pad and the sides of the trench were examined. Although the soil was well compacted, clod interfaces could be seen and poor bonding at lift interfaces was observed.

In hopes of improving performance, the test pad was reconstructed. The moisture content was increased by an average of 6%, enough to soften the clods resulting in their destruction upon compaction. Careful attention was paid to eliminating lift interfaces. A second infiltration test was performed. The results are also shown on Fig. 3. An average hydraulic conductivity of $5x10^{-8}$ cm/s was obtained. A change in the construction specifications and better QA/QC decreased the potential leakage rate through the test pad by a factor of 12.

Importance of Redundancy in Verification Testing

A test pad was constructed in an arid region using a soil with a PI = 20 and an optimum moisture content of 15%. The test pad was constructed in two 150 mm lifts to a relative density of 95% of standard proctor maximum density. A decision was made to use a compaction moisture content lower than normal, hoping to reduce the shrinkage potential of the soil. It was realized that this would increase the hydraulic conductivity of the soil. It was felt that in the long term, the leakage rate through the drier material would be less than that through material placed at a higher moisture content with eventual formation of shrinkage cracks. The compaction moisture content ranged from 0% to 3% dry of optimum moisture content.

Laboratory tests on the material yielded an average hydraulic conductivity of $2x10^{-8}$ cm/s. An infiltration test was performed using an SDRI. After 10 days of testing steady state was achieved and an hydraulic conductivity of $5x10^{-6}$ cm/sec was measured. The high value of hydraulic conductivity was thought to be the result of experimental errors or improper compaction of the test area. As a check, second test pad was constructed on top of the first and the SDRI test was repeated. Steady state infiltration was achieved after 10 days of testing at a hydraulic conductivity of $2x10^{-6}$ cm/s.

At this point, additional laboratory studies were conducted and it was found that optimum moisture for the test pad material was 19%, not 15% as thought earlier. It was unclear whether an error existed in the original determination of optimum moisture or if a material change had occurred. A third test pad was constructed, the compaction moisture content increased 7%, and a third SDRI test performed. After 65 days of testing, an hydraulic conductivity of $1x10^{-7}$ cm/s was achieved.

If an in situ hydraulic conductivity test had not been performed, leakage through the compacted material may have been 20 to 50 times what the designer had intended. It should be pointed out that with a QA/QC plan requiring the performance of field proctor tests, the change in optimum moisture content would have been detected earlier in the project.

Influence of Sample Size and Proper QA/QC

The following data are from two separate projects referred to as Case A and Case B. Test pads were constructed in both cases and verification testing consisted of laboratory tests using 75 mm and 300 mm diameter cylindrical samples and an SDRI test. Both soils had a high PI and were compacted to high relative densities. In Case A, the soil was compacted at optimum moisture content, and in Case B, the compaction moisture content was 4% greater than optimum.

In Case A, the soil was not pre-processed and inspection of the fill body indicated poor bonding at lift interfaces and numerous clod interfaces. In Case B, the material was pre-processed to reduce clod size, construction equipment and procedures were selected to minimize secondary features, and QA/QC was carefully enforced. The results of the hydraulic conductivity testing are summarized below.

		Hydraulic Conductivity, x10-7 cm/s		
		Lab	Lab	
	QA/QC	75 mm	300 mm	SDRI
Case A	poor	0.04	0.2	1
Case B	good	0.04	0.05	0.05

The benefits of a good QA/QC program are obvious as indicated by the excellent agreement obtained between the various test methods in Case B. Apparently there was no significant secondary structure in this soil. The problems that result from poor quality control are apparent from the lack of agreement between the various test methods used in Case A. The hydraulic conductivity in the field was 25 times greater than that determined in the laboratory on 75 mm diameter samples. The presence of secondary features were apparent in the results of the laboratory test on the 300 mm sample, however, the hydraulic conductivity was still five times less than that determined using an SDRI.

In all the case histories presented, suction at the wetting front was assumed to be zero when calculating the hydraulic conductivity from an SDRI test. This assumption is conservative and may result in a calculated hydraulic conductivity that is larger than the in situ hydraulic conductivity. This may account for part of the difference between the laboratory and in situ hydraulic conductivities presented above.

Performance Evaluation - Future

Increasing understanding of factors affecting performance of earthen liners has led to improvements in performance evaluation. This has also resulted in increased awareness of the limited amount of verification testing actually being performed. Presently, in the absence of full scale verification tests, much of the success of current liners relies on the QA/QC programs. It is important that those overseeing the construction of earthen liners realize the limitations of current performance testing. These limitations include:

1. lack of reliable verification testing during construction. (Large diameter laboratory tests may fill this gap.)

2. testing confined to a small flat test pad.
3. lack of field tests for testing critical sections of the liner (e.g. slopes, joints where slopes and bottom sections meet, and boundaries around sumps.)
4. lack of consideration of long term conditions which may include deformations in the liner, drying of the liner (especially in areas where liners are in contact with unsaturated soils)
5. the performance of tests with water, rather than leachates.

Current performance evaluation hinges on successful verification testing in the preconstruction phase. It is important that the QA/QC program ensure that the same care and attention used in constructing a test pad, be used during the construction phase. Success in passing preconstruction verification tests must not result in less attention to detail during the construction phase.

The authors would also like to point out that proper design should not be compromised in order to pass preconstruction verification testing. An example of this is liners used in arid regions. Lower hydraulic conductivity values can be achieved if a high compaction moisture content is used. However, this may also result in high shrinkage potential. If the liner is placed in contact with unsaturated soils, drying of the liner will result and if overburden is not significant, cracks will form through the liner. Better performance would probably be obtained by compacting the soil to a lower moisture content and higher density. This may result in higher hydraulic conductivity being measured in preconstruction verification testing, but would result in lower long term leakage rates.

As a final point, the authors would like to note that the trend to judge the suitability of a liner based on the results of a single type of verification test should be reconsidered. Limitations of current verification techniques make it necessary to consider other factors as well. For example, if a soil is considered marginal due to the results of an SDRI test, it may still perform as an adequate barrier to leakage if overburden stress is considered.

Summary and Conclusions

Performance evaluation techniques for earthen liners have changed significantly over the past decade. Up until recently, performing laboratory tests on small diameter samples was considered an appropriate means for assessing the performance of a full scale liner. Generally these tests were performed in the design phase and it was assumed that the field hydraulic conductivity would be the same as that determined in the laboratory.

Current performance evaluation consists of using a wide variety of testing methods in all phases of the liner project. Performance testing in preconstruction phase is used to refine soil specifications, evaluate compaction equipment and procedures to help formulate a QA/QC program. In the construction phase, performance testing helps to ensure that the QA/QC plan is working. And finally, performance testing in the service phase is used to check that the liner is performing as expected.

The case histories presented as well as other studies have demonstrated that secondary features can result in an in situ hydraulic conductivity that is greater than that determined only from laboratory tests using small diameter samples. Effective testing techniques need to account for secondary features if reliable results are to be expected. Secondary features are accounted for by permeating large volumes of soil.

Underdrains and infiltrometers are strongly recommended because of the capability to permeate large volumes. Underdrains have long testing times and are recommended for performance evaluation during the service phase. Infiltrometers are recommended for the preconstruction phase for use on testpads.

A shortcoming of the current performance evaluation process is the lack of a reliable verification test for the construction phase. The use of underdrains or infiltrometers is not compatible with construction schedules. The authors suggest that laboratory tests on large samples may be suitable for this purpose. Testing times are relatively short and sampling could be accomplished without interrupting construction. In one of the case histories presented, it was demonstrated that secondary effects were detected when testing 300 mm diameter samples, however not as well as with the SDRI.

It should be noted that even though the amount and quality of performance testing has significantly improved, it is still very limited. There is ample room for further advances. For instance, techniques are needed for testing critical sections of the liner such as slopes and interfaces around sumps.

Finally, while performance testing can help to improve the chances of building an effective earthen liner, the importance of having experienced field personnel to ensure that the QA/QC program is enforced cannot be overemphasized.

References

Bouwer, H.(1966). "Rapid Field Measurement of Air Entry Value and Hydraulic Conductivity of Soil as Significant Parameters in Flow System Analysis." Water Resour. Res.,2,729-732.

Daniel, D.E.(1984)."Predicting the Hydraulic Conductivity of Compacted Clay Liners." Journal of Geotechnical Engineering., ASCE, 110(2),285-300.

Daniel, D.E., Anderson, D. C., and S. S. Boyton (1985), "Fixed-Wall vs. Flexible-Wall Permeameters," ASTM STP 874, pp. 107-126.

Daniel, D.E., and Trautwein, S.J. (1986)."Field Permeability Test for Earthen Liners." Pro.,In Situ '86, Blacksburg, Va., 146-160.

Daniel, D.E., (1989) "In Situ Hydraulic Conductivity Test for Compacted Clay," Journal of Geotechnical Engineering Vol. 115, No. 9, pp. 1205-1226.

Day, S. R. and Daniel, D. E. (1985) "Hydraulic Conductivity of Two Prototype Clay Liners." Journal of Geotechnical Engineering, ASCE, Vol III, No. 8, August 1985, pp. 957-970.

Elsbury, B.R., Sraders, G.A.,Anderson,D.C., Rehage, J.A., Sai, J.O., and D. E. Daniel (1990), "Field and Laboratory Testing of a Compacted Soil Liner," U. S. EPA, Cincinnati, Ohio, EPA/600/S2-88/067.

Gordon, M. E., Huebner, P.M., and T. J. Miaza (1989). "Hydraulic Conductivity of Three Landfill Clay Liners," Journal of Geotechnical Engineering, Vol. 115, No.8, pp. 1148-1160.

Mitchell,J.K., Hooper, D.R., and Campanella, R.G. (1965). "Permeability of Compacted Clay." J. Soil Mech.Found. Div.,ASCE, 91(4), 41-65.

Reynolds, W.D., and Elrick,D.E. (1985)."In Situ Measurement of Field-Saturated Hydraulic Conductivity, Sorptivity, and the a-Parameter using the Guelph permeameter." Soil Sci 140(4), 292-302.

Rogowski,A.S. (1986)."Hydraulic Conductivity of Compacted Clay Soils" Proc., Twelth Annual Res. Symp. on Land Disposal, Remedial Action, Incineration, and Treatment of Hazardous Waste, U.S.EPA, Cincinnati, Ohio, 29-39.

Soilmoisture Equipment Corp. (1987).Model 2800 Guelph Permeameter. Soilmoisture Equipment Corp., Santa Barbara, Calif.

Soil Testing Engineers,Inc. (1983). STEI Two-Stage Permeability Test. Soil Testing Engineers, Inc., Baton Rouge, La.

Torstensson, B. A. (1984)."A New System for Ground Water Monitoring." Ground Water Monitoring Rev., 4(4), 131-138.

Trautwein Soil Testing Equipment Co. (1987). Installation and Operating Instructions for the Sealed Double Ring Infiltrometer. Trautwein Soil Testing Equipment Co., Houston, Tex.

U. S. Environmental Protection Agency (1986), "Technical Guidance Document : Construction Quality Assurance for Hazardous Waste Land Disposal Facilities," Office of Solid Waste, Washington, DC,EPA/530-SW-86-031.

FIELD BEHAVIOR OF DOUBLE-LINER SYSTEMS

Rudolph Bonaparte[1] A.M. and Beth A. Gross[1] A.M.

Abstract

This paper presents field data on flows of liquid from the leakage detection layers of double-liner systems at 30 landfill and surface impoundment facilities. For each facility, information on design and operation is presented, as is an evaluation of the sources of the measured flow. Potential sources include leakage through the top liner, precipitation that percolates into the leakage detection layer during construction, ground-water infiltration, and consolidation of any clay component of the top liner. From the evaluation, conclusions are drawn regarding the frequency of occurrence, sources, and rates of liquid flow from the leakage detection layers of double-liner systems. The evaluation also allows conclusions to be drawn on leakage rates through geomembrane liners, frequencies of defects in geomembrane liners, and proposed USEPA action leakage rates for hazardous waste storage and disposal facilities.

Introduction

Double-liner systems are being increasingly used to contain leachate and hazardous liquids in landfills and surface impoundments. The majority of double-liner systems being constructed today have either geomembrane or composite top and bottom liners (where a composite liner consists of a geomembrane placed directly on top of a low-permeability soil layer) with a leakage detection layer between the two liners (Figure 1). Rarely are low-permeability soil liners used alone in these systems.

[1]GeoSyntec Consultants (formerly GeoServices Inc. Consulting Engineers) 5950 Live Oak Parkway, Suite 330, Norcross, Georgia 30093

Liquid flows have been observed from the leakage detection layers of many double-lined landfill and surface impoundment facilities. Regulatory authorities and others often assume that these flows are due solely to leakage through the top liner and are, therefore, a cause for concern and action. For example, the U.S. Environmental Protection Agency (USEPA) has proposed regulations for hazardous waste disposal and containment facilities [i.e., U.S. Federal Register, Volume 52, Number 103, 20218-20309, 29 May 1987] which include a presumption that all flows in the leakage detection layer are due to leakage through the top liner. USEPA has proposed to set a threshold "action leakage rate" for these facilities in the range of 50 to 200 liters per hectare of lined area per day (lphd). If the average measured flow rate from the leakage detection layer exceeds the action leakage rate, the presumption is that excessive leakage through the top liner is occurring. In this case, USEPA will require the facility owner to provide a "response action plan" that describes the remedial actions that will be taken to mitigate the leakage. Only through a demonstration in the response action plan can the facility owner show that the flows are due to a source other than top liner leakage; with this demonstration, the owner may mitigate the need to implement remedial actions.

The gathering and analysis of data on flows from the leakage detection layers of double-lined landfills and surface impoundments is important for at least four reasons: (i) to establish the frequency of occurrence, sources, and rates of flow; (ii) to provide data and information to owners and engineers who must design, operate, and monitor the performance of double-liner systems; (iii) to assess the reasonableness of the range of action leakage rate values that have been proposed by the USEPA; and (iv) to develop data on top liner leakage rates so that the validity of the assumptions inherent in geomembrane liner leakage models, such as those presented by Bonaparte et al. [1989], Brown et al. [1987], Giroud and Bonaparte [1989a,b], and Jayawickrama et al. [1988], can be better assessed. With the exception of USEPA [1987] and Gross et al. [1990], there is currently no data in the technical literature on measured flow rates from the leakage detection layers of double-lined landfills and surface impoundments.

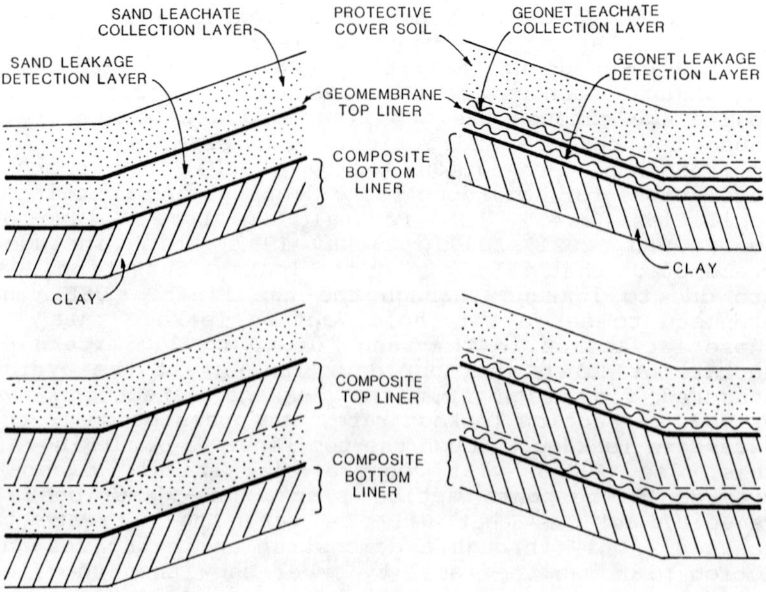

Figure 1. Typical double-liner systems for landfills.

Double-Liner Systems

Figure 1 illustrates typical double-liner systems used to contain leachate in landfills. Double-liner systems used to contain liquids in surface impoundments are similar to those shown in Figure 1, except that surface impoundments do not require a leachate collection layer above the top liner. (However, as discussed by Giroud and Bonaparte [1989a], if a composite top liner is used in a liquid impoundment, it should be covered with soil or other material to prevent its uplift in the event of rapid drawdown of the impoundment.) The double-liner systems shown in Figure 1 incorporate the liner types (i.e, geomembrane liners and composite liners) and drainage material types (i.e., granular materials and geonets) that will be considered in this paper. The types of liners and drainage materials significantly influence the frequencies of occurrence, sources, and rates of flow from leakage detection layers.

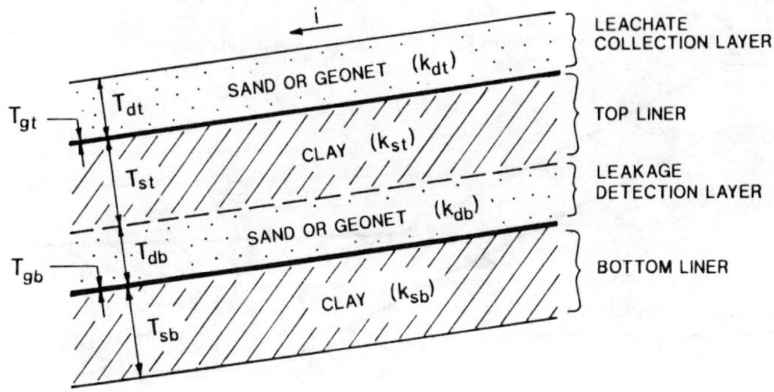

Figure 2. Definitions and terminology [Note: k = hydraulic conductivity; T = thickness].

In this paper, the parameters used to describe the design features of double-liner systems are (Figure 2): i = slope gradient (dimensionless); k_{dt} = hydraulic conductivity of the leachate collection layer (m/s); T_{dt} = thickness of the leachate collection layer (m); T_{gt} = thickness of the geomembrane component of the top liner (m); k_{st} = hydraulic conductivity of the clay component of the top liner (m/s); T_{st} = thickness of the clay component of the top liner (m); k_{db} = hydraulic conductivity of the leakage detection layer (m/s); T_{db} = thickness of the leakage detection layer (m); T_{gb} =thickness of the geomembrane component of the bottom liner (m); k_{sb} = hydraulic conductivity of the clay component of the bottom liner (m/s); and T_{sb} = thickness of the clay component of the bottom liner (m).

Potential Sources of Liquid Flow

To evaluate leakage detection layer flow data, the potential sources of flow must be identified. Flow can be due to [Gross et al., 1990] top liner leakage, water

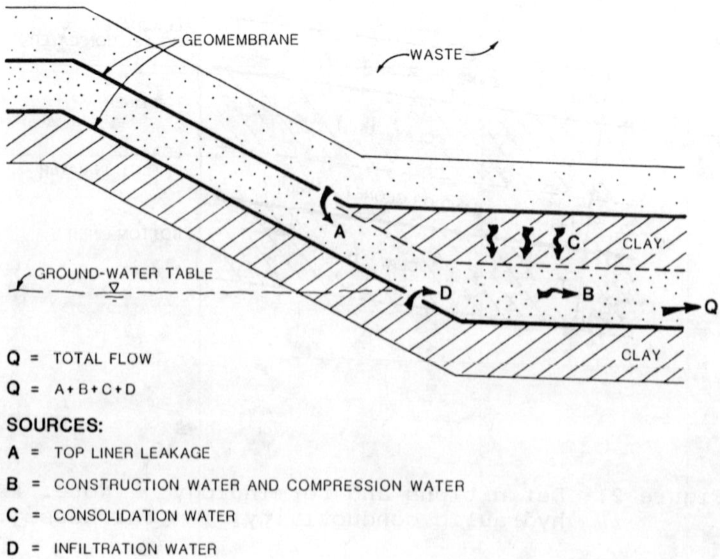

Figure 3. Sources of flow from leakage detection layers.

from precipitation that percolates into the leakage detection layer during construction ("construction water"), water expelled from the leakage detection layer due to compression of the layer ("compression water"), water squeezed out of the clay component of a composite top liner as a result of clay consolidation ("consolidation water"), and ground water that infiltrates the bottom liner and enters the leakage detection layer ("infiltration water"). Each of these potential sources of flow is depicted in Figure 3 and described below.

<u>Top Liner Leakage:</u> Leakage through liners constructed with geomembranes can occur as a result of flow throughholes in the geomembrane and permeation of the intact geomembrane. Equations to calculate steady-state leakage rates through liners constructed with geomembranes due to flow through holes and permeation through intact geomembranes have been presented by Bonaparte et al. [1989] and Giroud and Bonaparte

[1989a,b]. The equations show that the rate of permeation is a function of the type and thickness of the geomembrane, the chemical composition of the contained liquid, and the pressure difference acting across the geomembrane. The equations also show that the rate of flow through geomembrane holes is dependent on the liquid head acting on the geomembrane, the hydraulic conductivity of the soil layer immediately underlying the geomembrane, the size and frequency of occurrence of holes in the geomembrane, and for composite liners, the quality of the contact between the geomembrane and underlying soil layer (which is a function of the quality of construction).

Table 1 presents the results of calculations using the equations for flow through holes in the geomembrane and permeation of the intact geomembrane to obtain top liner leakage rates per unit area of liner. For the calculations, it was assumed that the permeating liquid was water, the geomembrane was a 1-mm thick high density polyethylene (HDPE) material, and the soil layer underlying the geomembrane was at least 1 m thick. It was further assumed that the geomembrane was carefully constructed and had only five small holes per hectare, with each hole having an area of 3×10^{-6} m^2, and with good quality contact between the geomembrane and underlying soil layer (from Giroud and Bonaparte [1989a,b]).

Table 1. Calculated top liner leakage rates due to water permeation and flow through holes in the geomembrane (liter/ hectare/day) [Note: k_{min} = min (k_{db} or k_{st})].

Hydraulic Conductivity k_{min} (m/s)	Liquid Head on Liner, h_t (m)		
	0.03	0.3	3.0
10^{-10}	0.05	1	80
10^{-8}	1	10	150
10^{-6}	40	300	2,500
10^{-4}	300	1500	5,600
10^{-2}	600	2000	6,000

From inspection of Table 1, it can be seen that calculated steady-state top liner leakage rates through composite liners (i.e., $k_{min} < 10^{-8}$ m/s) range from 0.05 to 150 liter/hectare/day (lphd). These calculated leakage rates are very low. In contrast, calculated leakage rates through geomembrane liners underlain by drainage materials having $k_{min} > 10^{-4}$ m/s range from 300 to 6,000 lphd. These leakage rates are comparatively high. A further contrast between composite liners and geomembrane liners lies in the fact that it can take from several months to many years for liquid to flow through the clay component of a composite liner, whereas flow through a hole in a geomembrane liner underlain by a drainage material occurs almost instantaneously.

<u>Construction Water:</u> The primary source of construction water in a leakage detection layer is precipitation that percolates into the layer prior to placement of the top liner. Of this water, some may be retained in the drainage material by capillary tension and the rest will flow by gravity from the layer. Gross et al. [1990] presented the results of calculations to quantify potential construction water flow rates and durations for a typical landfill cell. They found that the flow rate is directly proportional to k_{db} and the duration of flow is inversely proportional to k_{db}. Thus, with a high-permeability drainage material, such as a clean gravel or geonet (e.g., $k_{db} > 10^{-2}$ m/s), the flow rate after a precipitation event can be large (e.g., 200,000 lphd) but of very short duration (e.g., one day). In contrast, with a lower-permeability drainage material, such as a fine to medium sand (e.g., $k_{db} = 10^{-4}$ m/s), the rate of flow of construction water will be smaller (e.g., 2,000 lphd) but the duration of flow will be quite long (e.g., more than 100 days). In this latter case, construction water may still be draining from a facility well after the start of operation.

<u>Compression Water:</u> As a leakage detection layer constructed of a granular material compresses under the weight of the overlying waste or impounded liquid, the pore volume and porosity of the layer decrease. Simultaneously, the capillary tension of water in the pores of the material increases as the soil particles take on a denser packing. Calculations in Gross et al. [1990] indicate that the flow rate of compression water from a granular drainage material initially at field capacity is small (e.g., 1 to 20 lphd) in comparison to potential flow rates from other sources.

<u>Consolidation Water:</u> Two general categories of clay layers must be considered: (i) relatively thick (0.5 to 1.5 m thick) layers of compacted natural clay or bentonite-treated soil; and (ii) thin (6 mm thick) geotextile-bentonite mats. During filling of a landfill or surface impoundment, clay layers in the first category will consolidate and expel water into the adjacent

drainage layer. An upper bound of the rate at which water is expelled can be obtained by assuming the clay layer is initially saturated. With this assumption, the rate and occurrence of flow due to consolidation water can be quantified using the classical theory of one-dimensional consolidation [Terzaghi, 1943], which assumes an instantaneous load application, or Schiffman's [1957] modified theory of consolidation, which assumes a constant rate of load application.

Calculations presented in Gross et al. [1990] suggest that the flow rate due to consolidation water may range from 10 to 1,500 lphd. Higher flow rates are associated with more compressible soils, faster rates of load application, and larger consolidation stresses. The calculations also indicate that, for most landfills that receive waste at a steady rate over one or more years, the end of primary consolidation will approximately coincide with the end of the active life of the landfill. It should be noted however, that, water from secondary compression of a clay layer may persist after the end of primary consolidation. For some plastic clay layers compacted wet of optimum moisture content, water due to secondary compression may be a significant source of liquid flow after the end of the facility's active life. Calculated secondary compression flow rates are in the range of 10 to 100 lphd.

Infiltration Water: Ground water (i.e., infiltration water) can migrate through the bottom liner into the leakage detection layer if there is a sustained ground-water table above the base of the liner. The calculation results given in Table 1 can be used to estimate ground-water infiltration rates if it is assumed that the rate of flow through the liner is independent of the direction of flow (i.e., up or down). In this case, k_{min} in Table 1 corresponds to the hydraulic conductivity of the soil immediately underlying the geomembrane component of the bottom liner. From Table 1, the quantity of ground-water infiltration through composite bottom liners (i.e., $k_{sb} < 10^{-8}$ m/s) will be small and will only occur after ground water permeates the clay have to use "layer" layer. In contrast, ground-water infiltration rates through geomembrane bottom liners underlain by relatively permeable soils can be very large and will occur quickly.

Methodology for Evaluating Measured Flow

In the next section, the sources of flow from the leakage detection layer of operating facilities will be identified by comparing the measured flow rate for a specific time period with the calculated flow rates from different sources during the same time period. This methodology for evaluating the source of measured flow has been described by Gross et al. [1990] as follows:

1. Identify the potential sources of flow based on double-liner system design, climatic and hydrogeologic setting, and operating history.

2. Calculate flow rates from each potential source.

3. Calculate the time frame for flow from each potential source.

4. Evaluate the potential sources of flow by comparing measured flow rates to calculated flow rates at specific points in time.

Additionally, comparisons of the chemical constituents in liquids contained in surface impoundments or leachates from landfills with the chemical constituents in the flows from the leakage detection layer will provide insight into whether a source of the flow is leakage through the top liner.

<u>Data From Operating Facilities</u>

Leakage detection layer flow rate data have been collected from 23 double-lined landfills (containing 55 individually monitored landfill cells) and seven double-lined surface impoundments (containing 14 individually monitored ponds). The data, along with information on each of the facilities, are presented in Tables A-1 through A-6 at the end of this paper. Tables A-1 and A-2 present general information and double-liner system properties, respectively, for the landfill facilities. Tables A-4 and A-5 present the same information for the surface impoundment facilities. It is noted that the double-liner system properties given in Tables A-2 and A-5 are from project plans and specifications; they are not measured properties.

Tables A-3 and A-6, respectively, present the flow rate data for the landfill and surface impoundment facilities. Data are summarized for two different time periods: (i) just after the end of construction, when the influence of construction water would be greatest; and (ii) during the active life of the facility. Where possible, flow rates are reported at several different times during the active life. The authors found only limited data on flow rates after landfill closure. The data corresponding to closed landfill conditions are specially noted in Table A-3. For some facilities, liquids were pumped out of the leakage detection layers only periodically (e.g., several times a month) or flows were only recorded monthly. For these facilities, only an "average" flow rate is reported. Where there are sufficient data to evaluate the temporal variation in flow, both average and maximum flow rates are reported.

A characterization of the monitored cells and ponds is presented in Table 2. It can be seen that the monitored cells and ponds are almost equally divided between those with geomembrane top liners and those with composite top liners; they are also almost equally divided between those with sand leakage detection layers and those with geonet leakage detection layers. Most facilities are located at sites where the ground-water table is below the facility base. However, in a few cases, the depth to the ground-water table is unknown. In this paper, it will be assumed that ground-water infiltration did not contribute to the leakage detection layer flows at any of the facilities. For all facilities, the bottom liner included a geomembrane; in most cases, the bottom liner was a composite. Therefore, as shown in USEPA [1987], the collection efficiencies of the leakage detection layers at the monitored facilities should be very high (i.e., very little of the liquid in the leakage detection layer should have migrated into the bottom liner).

Table 2. Characterization of 69 individually monitored cells and ponds [Note: Gmb = geomembrane; Cmp = composite; Below = ground water below base facility].

	Top Liner		Leakage Detection Layer		Ground-Water Table	
	Gmb	Cmp	Sand	Geonet	Below	Unknown
Landfill	23	32	27	28	53	2
Impoundment	8	6	4	10	13	1

The authors have found it convenient to group the monitored facilities by the type of top liner (i.e., geomembrane versus composite) and type of leakage detection layer (i.e., sand versus geonet). With this grouping, the potential sources of flow for cells and ponds in any group are the same (Table 3).

Table 3. Number of facilities in each group.

Group No.	Top Liner	Leakage Detection Layer	Number of Cells and Ponds
I	Geomembrane	Geonet	13
II	Geomembrane	Sand	18
III	Composite	Geonet	25
IV	Composite	Sand	13

Group I Facilities

Group I facilities were constructed with geomembrane top liners and geonet leakage detection layers. Therefore, the only potential source of flow from the leakage detection layers of these facilities is top liner leakage (except for any small amount of construction water that drains from granular collection trenches in the leakage detection layer). All seven cells from the three Group I landfills appear to have exhibited top liner leakage, with average flow rates during the active lives of the cells ranging from 5 to 270 lphd. Maximum flow rates of up to about three times the average values were observed. The maximum flows typically corresponded to high flow rates from the leachate collection layers shortly after storm events.

The observed flow rates from the Group I landfills are, on average, somewhat smaller than the calculated top liner leakage rate (i.e., 600 lphd) given in Table 1 for $h_t = 0.03$ m and $k_{min} > 10^{-2}$ m/s. However, the calculated value appears to represent a reasonable upper bound of the observed values. It should also be noted that all of the Group I landfills are located in relatively moist climatic regions. All other things being equal, smaller rates of top liner leakage would be expected at facilities located in dry climates where leachate production rates are low.

There are two surface impoundments with six ponds in Group I. Four of the six ponds have exhibited no flow since the start of operation. Of the two ponds that have exhibited flow (i.e., Impoundment E, Pond 2 and Impoundment F), one exhibited an average flow of 30,000 lphd only after a 20-mm long defect developed along a geomembrane seam on the side slope of the pond. Flow has not been observed from this pond since the defect was repaired. The other pond initially exhibited an average flow of 400 lphd at 0.5 months after construction. This flow was due, in part, to top liner leakage. After a geomembrane defect was discovered and repaired, the flow rate decreased steadily with time and was at a very low value (i.e., 0.4 lphd) until 53 months after construction. At 53 months, the average flow rate increased to 4 lphd. Based on the increased flow rate and on testing of chemical constituents in the flow, it was determined that top liner leakage was occurring. The observed leakage coincided with an increase in the maximum liquid height in the pond. Therefore, the pond likely has a small geomembrane top liner defect on its side slope.

One reason for the absence of top liner leakage at four of the Group I ponds and only three occurrences of top liner leakage at the other two Group I ponds is that these facilities were all subjected to ponding tests and/or leak location surveys as part of the owner's internal or third-party construction quality assurance (CQA) program. Geomembrane defects identified during the leak location surveys and/or ponding tests were repaired.

It is useful to compare the observed top liner leakage rates at the monitored surface impoundments to the calculated leakage rates given in Table 1 for a liquid depth of 3.0 m. The calculated top liner leakage rates (which are obviously much larger than the observed flow rates from Group I surface impoundments reported in Table A-6) result from permeation of the intact geomembrane and flow through geomembrane holes. Giroud and Bonaparte [1989a,b] presented evidence suggesting that a geomembrane hole frequency of 2 to 5 holes per hectare and a geomembrane hole size of 3 x 10^{-6} m^2 may be assumed for calculating "representative" flow rates due to holes in geomembranes installed using rigorous CQA programs. It is apparent from the observations of the monitored facilities that the foregoing assumptions regarding the frequency and/or size of geomembrane holes are too conservative for geomembrane installations that include ponding tests and/or leak location surveys as part of a CQA program. The use of ponding tests and/or leak location surveys reduced the frequency and/or size of geomembrane holes in the Group I surface impoundments in this study below the values assumed by Giroud and Bonaparte. Their assumptions were based on analyses of the performance of geomembranes that had not been subjected to ponding tests and/or leak location surveys. In addition, in the model by Giroud and Bonaparte, the flow rate due to water permeation through an intact 1-mm thick HDPE geomembrane is calculated to be 80 lphd for a liquid head of 3.0 m. Since not even this quantity of flow was observed from the Group I surface impoundments in Table A-6, it is concluded that water permeation rates through HDPE geomembranes reported by Giroud and Bonaparte [1989a] for large liquid head conditions are too high.

Group II Facilities

Group II facilities were constructed with geomembrane top liners and sand leakage detection layers. There are seven landfills (16 cells) and two surface impoundments (two ponds) in this group, with all of the facilities located in relatively moist climatic regions. At the Group II facilities, k_{db} is in the range of 10^{-3} to 10^{-5} m/s. For this range of k_{db}, drainage of construction water could have occurred for about one week to one year after installation of the top liner. All of the landfill cells for which data exist (twelve cells) show significant leakage detection layer flows shortly after construction. The average measured flow rates at the end of construction range from 16 to 17,000 lphd, with four of the cells exhibiting average flow rates of 3,800 lphd or more. No end of construction flow data is available for the Group II surface impoundments.

Excluding Landfills J, V, and W (discussed subsequently), all nine cells from the Group II landfills exhibited flows during their active lives which are believed to be due to top liner leakage. The average flows from the nine cells potentially attributable to top liner leakage range from 2 to 3,300 lphd, with the maximum flows being up to ten times larger than the average values. Interestingly, the four cells at which CQA programs were not implemented (i.e., Landfills L and Q) had larger top liner leakage rates (i.e., 100 to 3,300 lphd) than the cells at which CQA programs were implemented (i.e., 2 to 350 lphd) (Table 4). The two Group II surface impoundment ponds had average flows potentially attributable to top liner leakage of from 27 to 80 lphd. Ponding tests were reportedly performed as part of the CQA program at the two ponds.

It is interesting to compare the flows attributable to top liner leakage for the Group I and II landfills (excluding Landfills J, V, and W) to USEPAs proposed action leakage rate, which ranges from 50 to 200 lphd. For this comparison, the 16 landfill cells have been subdivided into those that had a CQA program (eleven cells) and those that did not have a CQA program (five cells). From Table 4, it can be seen that of the eleven cells that had a CQA program, four had average flow rates less than 50 lphd. Eight of the eleven cells that had a CQA program exhibited flow rates of less than 200 lphd. On the basis of this limited available data, it would appear that an action leakage rate of 50 lphd is too low, since properly constructed geomembrane top liners that have undergone CQA monitoring cannot consistently limit top liner leakage to a value of less than 50 lphd. However, based on the data in Table 4, it appears that an action leakage rate of 200 lphd may be appropriate. Table 4 also shows that implementation of a CQA program during construction reduces top liner leakage rates in comparison to facilities that do not have a CQA program.

Table 4. Comparison of average flow rates at Group I and II landfills [Note: excludes Landfills J, V, and W].

Leakage Detection Layer Flow Rates	No. of Cells	
	CQA	No CQA
Flow rate less than 50 lphd	4	-
Flow rate in range of 50 to 200 lphd	4	1
Flow rate in range of 200 to 500 lphd	3	-
Flow rate greater than 500 lphd	-	4

Landfills V and W were not included in the comparison in Table 4. Both of these landfills were constructed in mid-1989 and both contain relatively low permeability leakage detection layers (i.e., $k_{db} \approx 10^{-5}$ m/s). Calculations suggest that drainage of construction water is still occurring at these facilities; the presence of construction water makes it nearly impossible to quantify the rate of any top liner leakage. It is noted, however, that top liner leakage is certainly occurring at Cell 1 of Landfill V. A slope failure occurred at this facility shortly after construction. The failure was confined to the liner system on the 12-m high, 3H:1V (horizontal: vertical) side slope. The failure involved the downslope sliding of the 0.6-m thick sand leachate collection layer and underlying geotextile cushion layer on the HDPE geomembrane top liner. Flow from the cell had not been monitored prior to the failure. Shortly after the failure, however, the flow rate from the leakage detection system was recorded at almost 1,400 lphd.

Landfill J is a special Group II facility because it was constructed with the compacted clay component of the bottom liner above, rather than below (as is normal), the geomembrane component. In addition, at Landfill J the leakage detection layer is continuous between cells, thereby allowing flow to cross from one cell to the next. A detailed analysis of leakage detection layer flows in Landfill J was presented in Gross et al. [1990]. Their analysis indicated construction water flow rates of 60 to 17,000 lphd. Consolidation water flow rates from the four Landfill J cells varied from less than 10 lphd for Cell 3, which was filled slowly over 18 months, to 1,700 lphd from Cell 1, which was filled in one month. Based on an analysis of chemical constituents in flow from the leakage detection layers and leachate collection layers, the top liners at two of the four Landfill J cells leaked (i.e., Cell 1 and Cell 4). In the case of Cell 1, no top liner leakage was observed until several years after closure, at which time flow from the leakage detection layer of 140 lphd commenced. An evaluation of the methods used to construct Cell 1 resulted in the conclusion that the most likely cause of top liner leakage after landfill closure was the development of a hole at a leachate collection pipe penetration of the geomembrane top liner.

Group III Facilities

Group III facilities were constructed with composite top liners and geonet leakage detection layers. For these facilities, top liner leakage rates should be very low and will not occur for some period after construction. Since geonets were used in the leakage detection layers of these facilities, there should not be any construction water (except for any small amount of water that drains from any granular collection trenches that drain the layer). Therefore, flows from the Group III facilities should result primarily from consolidation of the clay component of the top liner.

There are eight landfills containing 21 cells in the Group III category. With one exception, all of the cells exhibited flows from their leakage detection layers. Average flow rates during the active lives of these cells ranged from 20 to 840 lphd, with nine of the cells exhibiting flows greater than 200 lphd. The flows from these cells are primarily attributable to consolidation water and the flow rates are consistent with the calculated range of values of 10 to 1,500 lphd reported previously. For Group III cells that are closed (i.e., Landfill S, Cells 1 to 4 and Landfill T, Cells 1 and 2), ongoing leakage detection layer flows may be due to secondary compression of the clay component of the top liner, and, for Landfill S, possibly, infiltration water from consolidation of the saturated clay foundation that underlies the landfill; also, a small amount of the flow can be attributed to top liner leakage. Flows from the four Group III ponds are similar to those from the Group III landfill cells, with average consolidation water flow rates ranging from 14 to 230 lphd. It is also interesting to observe how the flow rate of consolidation water from a facility decreases over time (e.g., Landfill E, Cell 1; Surface Impoundment C, Pond 3).

Group IV Facilities

Group IV facilities were constructed with composite top liners and sand leakage detection layers. At the end of construction, flows from the leakage detection layers of these facilities should be due primarily to drainage of construction water. Subsequently, consolidation water will contribute to the flow from those facilities in which a conventional compacted clay layer was used in the composite top liner. However, in one of the facilities (Landfill I), a 6-mm thick prefabricated geotextile-bentonite mat was placed directly under the geomembrane. This mat is installed with the bentonite in a dry state. The bentonite hydrates in the presence of water, thereby forming a low-permeability barrier. Based on the results of consolidation tests, the quantity of consolidation water squeezed out of a prefabricated geotextile-bentonite mat during filling of a landfill will be on the order of 1 to 10 lphd, if the bentonite is initially saturated. If the bentonite is not initially saturated, the quantity of consolidation water will be smaller.

There are eleven landfill cells in Group IV. Four of the cells have a layer of compacted clay as the soil component of the composite top liner and seven have a geotextile-bentonite mat. Construction water flow rates at these facilities ranged from 15 to 23,000 lphd. Consolidation water flow rates from the four cells with a layer of compacted clay in the composite top liner ranged from 20 to 189 lphd, which is consistent with the consolidation water flow rates observed for the Group III facilities.

In two of the seven cells with geotextile-bentonite mats, no leakage detection layer flow was observed. For the five remaining cells, an average flow of 90 lphd or less was observed during the cells' active lives. These average flow rates are low and could be accounted for by consolidation of the geotextile-bentonite mat, compression and continuing drainage of the sand leakage detection layer, or leakage through the geomembrane top liner on the cells' side slopes (the geotextile-bentonite mat only extends over the base of the cells).

There is one surface impoundment in Group IV and it contains two ponds. Both ponds were constructed using compacted clay as the soil component of the composite top liner. Flow rates from one of the ponds at 30 to 48 months after construction generally ranged from 100 to 200 lphd. Given the timeframe for this flow, consolidation of the 0.5-m thick layer of compacted clay should be nearly complete. The calculated breakthrough time for flow through the compacted clay component of the composite top liner of this cell is less than one year. Consequently, some of the flow from this pond may be due to top liner leakage. For this same pond, the flow rate from the leakage detection layer increased from about 200 to 1,200 lphd at 36 months after construction. The increase coincided with an increase in the liquid level in the pond. Based on testing of chemical constituents in the flow, it was determined that top liner leakage was occurring. The leakage detection layer flow rate decreased when the liquid level was lowered. It therefore appears that the geomembrane top liner on the pond side slope had a defect (the clay component of the composite top liner did not extend all the way up the side slope). The flow rate from the leakage detection layer of the second pond was zero from 20 to 43 months after construction; for this latter pond, consolidation water flows apparently ceased prior to flow rate measurement and quantifiable top liner leakage did not occur.

As was done with the Group I and II landfills, the flows from the leakage detection layers of the Group III and IV landfills can be compared to USEPAs proposed action leakage rate (Table 5). In the case of Group I and II landfills, flows from the leakage detection layers are due primarily to top liner leakage. For Group III and IV landfills, flows are primarily due to consolidation water. While consolidation water is not an environmental concern, USEPA intends to consider all flow from the leakage detection layer as top liner leakage unless it is demonstrated (in a response action plan) to be from another source. For this reason, it is interesting to see if consolidation water flow rates are above or below USEPAs proposed action leakage rate.

Table 5. Comparison of average flow rates at Group III and IV landfills [Note: Excludes Landfills I and S].

Leakage Detection Layer Flow Rates	No. of Cells
Flow rate less than 50 lphd	5
Flow rate in range of 50 to 200 lphd	8
Flow rate in range of 200 to 500 lphd	3
Flow rate greater than 500 lphd	2

Landfill I was excluded from Table 5 because the composite top liner at this facility has a geotextile-bentonite mat rather than a layer of compacted clay. Landfill S was excluded because it is suspected that some of the observed flow is due to infiltration of water from consolidation of the saturated clay foundation that underlies the landfill. From Table 5, it can be seen that only five of the 18 Group III and IV landfill cells for which data exist have consolidation water flow rates of less than 50 lphd, whereas 13 of the 18 cells have consolidation water flow rates less than 200 lphd. On this basis, it would appear that an action leakage rate of 50 lphd is too restrictive and a value of 200 lphd may better reflect an achievable performance level. This conclusion should be interpreted cautiously, however, because consolidation water flow rates vary with time and the reported values may not be maximums.

Conclusions

From the data presented in this paper, the following conclusions are drawn.

- Leakage detection layers in double-liner systems frequently exhibit flows that may be due to leakage through the top liner or to other sources such as construction water, consolidation water, and infiltration water. Leakage detection layer flow rate data presented in Tables A-3 and A-6, for landfills and surface impoundments, respectively, demonstrate the frequencies of occurrence and rates of flows from these sources.

- All of the double-lined landfill cells reviewed in this study that were constructed with geomembrane top liners appear to have exhibited top liner leakage. Based on the available data, the flow rates attributable to top liner leakage at active cells that had geomembrane top liners and CQA programs were frequently less than 200 lphd; the maximum measured flow rates, which were often associated with increased flow from the leachate collection layers shortly after storm events, were typically several times the average flow rates.

- Very little leakage detection layer flow was observed at double-lined surface impoundment ponds constructed with geomembrane top liners. The low flows may be attributed to the use of ponding tests and/or leak location surveys to identify geomembrane defects and allow their repair.

- The leakage calculation method presented by Giroud and Bonaparte [1989a,b] provided a reasonable upper bound of the observed flow rates attributable to top liner leakage at the Group I and II landfills. However, the method greatly overpredicted top liner leakage rates at the Group I and II surface impoundments. It appears that the primary reason for the overprediction is that the number and/or frequency of geomembrane holes assumed by Giroud and Bonaparte are too high. The surface impoundments described in this study were subjected to ponding tests and/or leak location surveys as part of their CQA programs. The use of ponding tests and/or leak location surveys reduces the frequency and/or size of geomembrane holes below those assumed by Giroud and Bonaparte. Furthermore, the absence of any flow from most of the Group I and II surface impoundments suggests that the rates of water permeation through intact geomembranes reported by Giroud and Bonaparte [1989a] for large liquid head conditions are too high.

- The double-lined landfills and surface impoundments in this study having a layer of compacted clay as the soil component of a composite top liner almost always exhibited flows due to consolidation water. Measured flow rates attributable to consolidation water were in the range of 20 to 840 lphd. Only very small flows were observed from the leakage detection layers of cells where the soil component of the composite top liner was a prefabricated geotextile-bentonite mat.

- The calculation methods presented by Gross et al. [1990] for estimating consolidation water and construction water flow rates appear reasonable for the facilities reported in this study.

- Based on the data in this study, an action leakage rate of 50 lphd is too restrictive and presents a performance standard that, if promulgated by USEPA, frequently will not be met by facilities that were constructed to present standards with rigorous third-party CQA programs. An action leakage rate of 200 lphd appears to be reasonable for landfills that have been constructed using rigorous third-party CQA programs. Even at this level, the action leakage rate may be temporarily exceeded at the start of operation of a facility, due to drainage of construction water, and, for facilities with composite top liners, during the active life of the facility, due to drainage of consolidation water. For surface impoundments with geomembrane top liners, ponding tests and/or leak location surveys will be required to meet a leakage detection layer performance standard of 200 lphd.

- The data presented in Tables A-3 and A-6 suggest that the double-liner systems evaluated in this study have performed well. Leakage rates through the top liners have been low or negligible in most cases and the leakage detection layers appear to be functioning effectively.

Acknowledgements

The authors extends their thanks to all of the facility owners and operators and to the regulatory officials who contributed data to this paper, to Alessi D. Otte of the U.S. Environmental Protection Agency, Office of Solid Waste, who early on recognized the need to obtain data on double-liner system performance, to Dr. Jean-Pierre Giroud, who reviewed the paper, and to Michelle Szalkowski, who typed the paper.

References

Bonaparte, R., Giroud, J.P., and Gross, B.A., "Rates of Leakage Through Landfill Liners", Proceedings, Geosynthetics '89, San Diego, Feb 1989, Vol. 1, pp. 18-29.

Brown, K.W., Thomas, J.C., Lytton, R.L., Jayawickrama, P., and Bahrt, S.C., "Quantification of Leak Rates Through Holes in Landfill Liners", USEPA Report CR 810940, Cincinnati, 1987, 147 p.

Giroud, J.P. and Bonaparte, R., "Leakage Through Liners Constructed with Geomembranes - Part I. Geomembrane Liners", Geotextiles and Geomembranes, Vol. 8, No. 1, 1989a, pp. 27-67.

Giroud, J.P. and Bonaparte, R., "Leakage Through Liners Constructed with Geomembranes - Part II. Composite Liners", Geotextiles and Geomembranes, Vol. 8, No. 2, 1989b, pp. 78-111.

Gross, B.A., Bonaparte, R., and Giroud, J.P., "Evaluation of Flow from Landfill Leakage Detection Layers", Proceedings, Fourth International Conference on Geotextiles, Vol. 2, The Hague, Jun 1990, pp. 481-486.

Jayawickrama, P., Brown, K.W., Thomas, J.C., and Lytton, R.L., "Leakage Rates Through Flaws in Geomembrane Liners", Journal of Environmental Engineering, ASCE, Vol. 114, No. 6, Dec 1988, pp. 1401-1420.

Schiffman, R.L., "Consolidation of Soil Under Time Dependent Loading and Varying Permeability", Highway Research Board Proceedings No. 37, 1958, pp. 584-617.

Terzaghi, K., Theoretical Soil Mechanics, John Wiley and Sons, Inc. New York, 1943, 527 p.

USEPA, "Background Document: Proposed Liner and Leak Detection Rule", EPA\530-SW-87-015, May 1987, 526 p.

Table A-1. General information on double-lined landfills.

Landfill (Group)	Location (U.S.)	Average Annual Rainfall (mm)	Ground Water Separation Distance (m)	Area of Cell (hectares)	Maximum Waste Height (m)	Drain Collector Spacing (m)	Base Slope Gradient	End of Construction	Third-Party CQA of Gmb Installation	Method of Flow Rate Measurement
A (IV)	Unknown	Unknown	Unknown	3.6	Unknown	Unknown	Unknown	5/1986	Yes	Weekly pumping if liquid present in sump, liquid volume recorded from accumulating flow meter
B (IV)	Unknown	Unknown	Unknown	4.5	Unknown	10	Unknown	3/1986	Yes	Weekly pumping if liquid present in sump, liquid volume recorded from accumulating flow meter
C (III)	Unknown	Unknown	1.5	1.2	Unknown	Unknown	Unknown	9/1985	Yes	Liquids pumped from sump, liquid volume recorded daily
D (IV)	Unknown	Unknown	6	1.1	Unknown	Unknown	Unknown	1986	Unknown	Unknown
E (III) - Cell 1	North Central	790	3	3.3	20	30	0.02	9/1986	No	Periodic pumping if liquid present in sump, liquid volume recorded monthly from accumulating flow meter
Cell 2				2.7		30	0.02	10/1987		
Cell 3				2.3		30	0.02	6/1988		
Cell 4				2.1		30	0.02	3/1990		
F (III)	North East	970	0	1.3	Unknown	40	0.03	8/1986	Yes	Periodic pumping if liquid present in sump, liquid volume recorded monthly from accumulating flow meter
G (III)	North East	970	0	2.7	Unknown	40	0.03	2/1987	Yes	Weekly pumping with manually operated electric pumps, liquid volume recorded from accumulating flow meter
H (III) - Cell 1	South West	280	90	4.6	30	30	0.02	3/1987	Yes	Daily pumping with jet pump if liquid depth in sump > 0.3 m, liquid volume recorded from accumulating flow meter
Cell 2				4.3	30	30	0.02	12/1987		
Cell 3				4.6	30	30	0.02	6/1987		
Cell 4				3.8	30	30	0.02	7/1989		
I (IV) - Cell 1	North East	1,040	1	2.0	25	30	0.02	7/1988	Yes	Daily pumping if liquid present in sump, liquid volume recorded from accumulating flow meter
Cell 2				2.0	25	30	0.02	7/1988		
Cell 3				1.7	25	30	0.02	8/1988		
Cell 4				1.7	25	30	0.02	8/1988		
Cell 5				2.8	25	30	0.02	9/1988		
Cell 6				3.9	25	30	0.02	11/1988		
Cell 7				2.8	25	30	0.02	1/1989		
J (II) - Cell 1	Gulf Coast	1,630	2	0.4	8	37	0.02	9/1985	Yes	Liquid volume from sump measured and recorded weekly by opening drain lines and allowing gravity flow into a graduated container
Cell 2				0.3	8	37	0.02	9/1985		
Cell 3				0.2	8	37	0.02	6/1987		
Cell 4				0.4	8	37	0.02	6/1987		
K (I)	North Central	790	3	5.6	20	30	0.02	8/1985	No	Periodic pumping if liquids present in sump, liquid volume measured indirectly by multiplying pumping time by pump capacity
L (II) - Cell 1	North East	1,040	5	2.4	40	45	0.013	3/1988	No	Automatic pumping of liquid from sump, liquid volume measured indirectly by multiplying pumping time by bump capacity
Cell 2				2.4	40	45	0.013	10/1987		

Table A-1. General information on double-lined landfills. - (continued)

Landfill (Group)	Location (U.S.)	Average Annual Rainfall (mm)	Ground Water Separation Distance (m)	Area of Cell (hectares)	Maximum Waste Height (m)	Drain Collector Spacing (m)	Base Slope Gradient	End of Construction	Third-Party CQA of Gmb Installation	Method of Flow Rate Measurement
M (I) - Cell 1 Cell 2	South East	1,520	0.6	4.3 4.9	8 8	15 15	0.02 0.02	8/1988 2/1989	Yes	Liquid volume pumped from sump recorded periodically from accumulating flow meter
N (I) - Cell 1 Cell 2 Cell 3 Cell 4	North East	1,070	0.9	1.9 2.6 1.5 1.9	25 25 25 25	20 45 Unknown Unknown	0.027 0.04 Unknown Unknown	7/1986 1/1987 11/1987 3/1988	Yes	Flow rate measurements to sump taken continuously with automated flow monitoring system and weekly using a 0.9 l container and stopwatch
O (II) - Cell 1 Cell 2 Cell 3	North East	990	1.2	2.7 2.3 1.9	5 8 8	30 30 30	0.02 0.02 0.02	8/1987 10/1987 4/1988	Yes	Flow rate measurements to sump taken continuously with automated flow monitoring system
P (II) - Cell 1 Cell 2	North East	1,040	1.5 1.5	4.0 4.0	50 50	5 5	0.005 0.005	2/1987 2/1987	Yes	Automatic pumping of liquids from sump if liquid depth in sump > 1.5 m, liquid volume recorded from accumulating flow meter
Q (II) - Cell 1 Cell 2	North East	1,040	1.5 1.5	3.3 3.5	20 20	40 40	0.005 0.005	12/1985	No	Liquid volume pumped from sump estimated by multiplying the sump area by the change in liquid depth in the sump
R (IV)	North East	1,040	1.5	6.4	25	30	0.02	9/1987	No	Liquid volume pumped from sump recorded from accumulating flow meter
S (III) - Cell 1 Cell 2 Cell 3 Cell 4 Cell 5	North East	970	0	0.6 0.5 0.5 0.6 0.9	20 20 20 20 20	60 60 60 60 60	0.02 0.02 0.02 0.02 0.02	3/1986 3/1986 8/1986 8/1986 11/1988	Yes	Liquid volume pumped from sump recorded from accumulating flow meter
T (III) - Cell 1 Cell 2 Cell 3 Cell 4	South East	1,350	9	0.6 0.6 1.1 1.1	20 20 25 25	30 30 100 100	0.02 0.02 0.025 0.025	5/1985 7/1985 9/1989 9/1989	Yes	Liquid volume pumped from sump recorded from accumulating flow meter
U (III)	Gulf Coast	1,420	0.6	1.0	9	Unknown	0.03	12/1988	Yes	Liquid volume pumped from sump recorded from accumulating flow meter
V (II) - Cell 1 Cell 2	North East	910	0.9	3.0 1.6	Unknown	45	0.02	5/1989	Yes	Flow rate measurements taken with pail and stopwatch
W (II)	North East	1,100	7.6	2.7	30	25	0.02	9/1989	Yes	Liquid volume pumped from sump recorded from accumulating flow meter

Table A-2. Details of landfill double-liner systems[(a)].

LANDFILL (GROUP) AND LINING SYSTEM DESCRIPTION	LEACHATE COLLECTION SYSTEM [(b)]			TOP LINER [(b)]				LEAKAGE DETECTION LAYER [(b)]			BOTTOM LINER (base slope) [(b)]			
	Material	k_{ct} (m/s)	D_{ct} (m)	k_{st} (m/s)	D_{st} (m)	Polymer	T_{gt} (mm)	Material	k_{cb} (m/s)	D_{cb} (m)	k_{sb} (m/s)	D_{sb} (m)	Polymer	T_{gb} (mm)
A (IV) composite/composite	Sand	10^{-4}	0.3	2×10^{-10}	0.6	HDPE	1.5	Sand	10^{-4}	0.3	2×10^{-10}	0.9	HDPE	1.5
B (IV) composite/composite	Sand	10^{-4}	0.3	2×10^{-10}	0.5	HDPE	1.5	Sand	10^{-4}	0.3	2×10^{-10}	0.9	HDPE	1.5
C (III) composite/composite	Geonet	10^{-3}	9×10^{-3}	10^{-9}	0.6	HDPE	1.5	Geonet	$>10^{-2}$	9×10^{-3}	10^{-9}	0.9	HDPE	1.5
D (IV) composite/composite	Sand	10^{-5}	0.3	10^{-9}	0.3	HDPE	2	Sand	10^{-5}	0.3	10^{-9}	0.9	HDPE	2
E[(c)] (III) composite/composite	Sand	2×10^{-4}	0.3	10^{-10}	1.5	HDPE	2	Geonet	$>10^{-2}$	1×10^{-2}	10^{-10}	3	HDPE	1.5[(c)] 2[(c)]
F (III) composite/composite	Gravel Overlying Geonet	10^{-2} $>10^{-2}$	0.3 6×10^{-3}	10^{-9}	0.5	HDPE	2	Sand Overlying Geonet	10^{-4} $>10^{-2}$	0.3 6×10^{-3}	10^{-9}	1	HDPE	1.5
G (III) composite/composite	Gravel Overlying Geonet	10^{-2} $>10^{-2}$	0.3 6×10^{-3}	10^{-9}	0.5	HDPE	2	Gravel Overlying Geonet	10^{-2} $>10^{-2}$	0.3 6×10^{-3}	10^{-9}	1	HDPE	2
H (III) composite/composite	Gravel Overlying Geonet	5×10^{-3} $>10^{-2}$	0.3 6×10^{-3}	10^{-9}	0.5	HDPE	1.5	Gravel Overlying Geonet	5×10^{-3} $>10^{-2}$	0.3 6×10^{-3}	10^{-9}	1	HDPE	1.5
I[(d)] (IV) composite/composite	Sand	10^{-4}	0.6	10^{-11}	6×10^{-3}	HDPE	1.5	Sand	10^{-4}	0.3	10^{-11}	0.2	HDPE	1.5
J[(e,f)] (II) geomembrane/composite	Sand	10^{-4}	0.3	--	--	HDPE	2	Sand	10^{-4}	0.3	10^{-9}	1	HDPE	1[(f)] 1.5[(f)]
K (I) geomembrane/composite	Sand	4×10^{-5}	0.3	--	--	HDPE	1.5	Geonet	$>10^{-2}$	6×10^{-3}	10^{-10}	5	HDPE	1.5
L (II) geomembrane/composite	Sand	10^{-5}	0.6	--	--	CSPE	0.9	Sand	$>10^{-5}$	0.6	10^{-5}	0.6	PVC	0.75
M (I) geomembrane/composite	Sand	10^{-5}	0.3	--	--	HDPE	2	Sand Overlying Geonet	10^{-5} $>10^{-2}$	0.3 0.3	10^{-8}	0.6	HDPE	1
N (I) geomembrane/composite	Sand	10^{-4}	0.6 - 0.9	--	--	HDPE	1.5	Geonet	$>10^{-2}$	6×10^{-3}	10^{-7}	0.3	HDPE	1.5

Table A-2. Details of landfill double-liner systems[(a)]. - (continued)

LANDFILL (GROUP) AND LINING SYSTEM DESCRIPTION	LEACHATE COLLECTION LAYER [(b)]			TOP LINER [(b)]				LEAKAGE DETECTION LAYER [(b)]			BOTTOM LINER (base slope) [(b)]			
	Material	k_{dt} (m/s)	D_{dt} (m)	k_{st} (m/s)	D_{st} (m)	Polymer	T_{gt} (mm)	Material	k_{db} (m/s)	D_{db} (m)	k_{sb} (m/s)	D_{sb} (m)	Polymer	T_{gb} (mm)
O (II) geomembrane/composite	Sand	10^{-3}	0.6	--	--	HDPE	1.5	Sand	10^{-3}	0.4	10^{-7}	0.3	HDPE	1.5
P (II) geomembrane composite	Sand	10^{-5}	0.6	--	--	CSPE	0.9	Sand	10^{-5}	0.6	10^{-9}	0.6	PVC	0.75
Q (II) geomembrane/geomembrane	Sand	10^{-4}	0.5	--	--	CSPE	0.9	Sand	10^{-4}	0.5	10^{-5}	0.2	PVC	0.75
R (IV) composite/composite	Sand	10^{-4}	0.5	10^{-9}	0.6	CSPE	0.9	Sand	10^{-4}	0.5	10^{-9}	0.2	PVC	0.75
S (III) composite/composite	Sand Overlying Geonet	10^{-4} $>10^{-2}$	0.3 6×10^{-3}	10^{-9}	0.5	HDPE	2	Sand Overlying Geonet	10^{-4} $>10^{-2}$	0.3 6×10^{-3}	10^{-9}	0.6	HDPE	1.5
T (III) composite/composite	Sand Overlying Geonet	10^{-4} $>10^{-2}$	0.1 6×10^{-3}	10^{-9}	0.9	HDPE	1.5	Geonet	$>10^{-2}$	6×10^{-3}	10^{-9}	0.9	HDPE	1.5
U (III) composite/composite	Geonet	$>10^{-2}$	1×10^{-2}	10^{-9}	0.9	HDPE	2	Geonet	$>10^{-2}$	1×10^{-2}	10^{-9}	0.9	HDPE	2
V (II) geomembrane/composite	Sand	5×10^{-5}	0.6	--	--	HDPE	1.5	Sand	5×10^{-5}	0.3	10^{-10}	0.6	HDPE	1.5
W (II) geomembrane/composite	Sand	10^{-5}	0.6	--	--	HDPE	2	Sand	10^{-5}	0.3	10^{-9}	0.6	LLDPE	1.5

Notes:

(a) The reported properties were obtained from engineering drawings, project specifications, and landfill owners and operators. If the properties of a project-specific geonet drainage layer could not be determined, the geonet was assumed to have a hydraulic conductivity greater than 10^{-2} m/s and a thickness of 6×10^{-3} m.

(b) Abbreviations are as follows: k_{dt} = hydraulic conductivity of leachate collection layer drainage material; D_{dt} = thickness of leachate collection layer drainage material; k_{st} = hydraulic conductivity of compacted clay component of top liner; D_{st} = thickness of compacted clay component of top liner; T_{gt} = thickness of geomembrane component of top liner; k_{db} = hydraulic conductivity of leakage detection layer drainage material; D_{db} = thickness of leakage detection layer drainage material; k_{sb} = hydraulic conductivity of compacted clay component of bottom liner; D_{sb} = thickness of compacted clay component of bottom liner; and T_{sb} = thickness of geomembrane component of bottom liner.

(c) The geomembrane component of the bottom liner is 1.5 mm thick for Cells 1 through 3 and 2 mm thick for Cell 4.

(d) The soil component of the composite top liner is a geotextile-bentonite mat.

(e) For this lining system, the order of the components of the bottom composite liner is reversed (the compacted low-permeability soil component of the bottom liner was placed on top of the geomembrane component) from the order normally used for a composite liner.

(f) The geomembrane component of the bottom liner is 1 mm thick for Cells 1 and 2 and 1.5 mm thick for Cells 3 and 4.

Table A-3. Measured flow rates from the leakage detection layers of the double-lined landfills.

Landfill (Group)	QUANTITY OF LIQUIDS MEASURED: END OF CONSTRUCTION: Time[a] (months)	Average Flow (lphd)	Maximum Flow (lphd)	Potential Sources of Flow	DURING ACTIVE LIFE: Time[a] (months)	Average Flow (lphd)	Maximum Flow (lphd)	Potential Sources of Flow
A (IV)	-	-	-	-	11	35	(b)	Water from consolidation of clay component of top liner
B (IV)	1	15	(b)	Construction water draining from sand	-	-	-	
C (III)	-	-	-	-	6	107	(b)	Water from consolidation of clay component of top liner
D (IV)	-	-	-	-	8	189	(b)	Water from consolidation of clay component of top liner and construction water draining from sand
E (III) - Cell 1	0.5	1,700	6,400	Construction water draining from granular collection trenches of Cells 1 and 4, and, for Cell 1 which is being loaded with waste, water from consolidation of clay component of top liner	3 36	840 100	(b) (b)	Water from consolidation of clay component of top liner for Cells 1 through 3
Cell 2	-	-	-		24	110	(b)	
Cell 3	-	-	-		18	510	(b)	
Cell 4	0.5	1,900	(c)		-	-	-	
F (III)	2	1,600	(c)	Construction water draining from granular collection trenches and water from consolidation of clay component of top liner	19	380	(c)	Water from consolidation of clay component of top liner
G (III)	2	130	(c)	Construction water draining from granular collection trenches and water from consolidation of clay component of top liner	10	20	(c)	Water from consolidation of clay component of top liner
H (III) - Cell 1	0.5	500	(c)	Construction water draining from granular collection trenches of Cells 1 through 3	12 24(e) 36(e)	80 55 65	(b) (b) (b)	Water from consolidation of clay component of top liner for Cell 1 at 12 months after construction and for Cells 2 through 4 (i.e., for active cells); water from secondary compression of clay component of top liner and/or top liner leakage for Cell 1 at 24 months and 36 months (after Cell 1 was covered with a geomembrane)
Cell 2	2	1	(c)		12 24	69 54	(b) (b)	
Cell 3	1	140	(c)		9	89	(c)	
Cell 4	1	0	0		9	0	0	
I (IV) - Cell 1	-	-	-	Construction water draining from sand for Cells 3 through 5	2 to 10	0	0	For Cells 3 through 7, possible sources of flow include construction water draining from sand, compression water draining from sand, leakage through defects in the top liner on base slope (i.e., defect in the geomembrane and geotextile-bentonite mat), and leakage through defect in geomembrane top liner on side slope (geotextile-bentonite mat does not extend up the side slope)
Cell 2	-	-	-		2 to 10	0	0	
Cell 3	0.5	200	650		2 to 3 4 to 8 8 to 9	10 0 90	(c) (c) (c)	
Cell 4	0.5	1,100	6,000		2 to 7 8 to 9	40 0	(c) (c)	
Cell 5	0.5	70	550		2 to 8	50	(c)	
Cell 6	-	-	-		2 to 6	50	(c)	
Cell 7	-	-	-		2 to 4	50	(c)	

Table A-3. Measured flow rates from the leakage detection layers of the double-lined landfills. - (continued)

Landfill (Group)	QUANTITY OF LIQUIDS MEASURED: END OF CONSTRUCTION				DURING ACTIVE LIFE			
	Time[a] (months)	Average Flow (lphd)	Maximum Flow (lphd)	Potential Sources of Flow	Time[a] (months)	Average Flow (lphd)	Maximum Flow (lphd)	Potential Sources of Flow
J (II) - Cell 1	1	17,000	32,000	Construction water draining from sand for Cells 1 through 4	3	1,700	4,300	Water from consolidation of clay component of bottom liner for all active cells experiencing flow; for Cell 1 after closure, top liner leakage; for Cell 4, top liner leakage; leakage determination based on testing chemical constituents in flow
					9 to 45	0	0	
					46(f)	20(f)	30(f)	
					52(f)	140(f)	200(f)	
Cell 2	1	4,700	17,000		6 to 16	0	0	
					24	110	130	
					36(f)	6(f)	10(f)	
					40 to 52(f)	0(f)	0(f)	
Cell 3	1	200	620		6	2	6	
					12	6	7	
					30	270	320	
Cell 4	1	60	470		6 to 23	0	0	
					24	120	1,300	
					30	270	320	
K (I)	-	-	-	-	10	57	(c)	Top liner leakage
L (II) - Cell 1	2	3,800	(b)	Construction water draining from sand for Cell 1 and 2	8	730	(b)	Top liner leakage for Cells 1 and 2; the higher leakage detection layer flow rates correspond with higher flow rates from the leachate collection layer associated with storm events
					12	460	(b)	
					16	3,300	(b)	
					20	2,050	(b)	
					24	880	(b)	
Cell 2	0.5	6,500	(c)		9	130	(b)	
					17	240	(b)	
					21	2,600	(b)	
					25	1,300	(b)	
					29	570	(b)	
					36	1,350	(b)	
M (I) - Cell 1	0.5	700	(c)	Construction water draining from granular collection trenches	4	130	(c)	Top liner leakage for Cells 1 and 2; for Cell 2 at 2 months, potential sources of flow also include construction water draining from granular collection trenches
					16	29	(b)	
					21	7	(b)	
Cell 2	-	-	-		2	260	(b)	
					6	56	(b)	
					10	23	(b)	
					15	5	(b)	

Table A-3. Measured flow rates from the leakage detection layers of the double-lined landfills. - (continued)

Landfill (Group)	QUANTITY OF LIQUIDS MEASURED							
	END OF CONSTRUCTION				DURING ACTIVE LIFE			
	Time[a] (months)	Average Flow (lphd)	Maximum Flow (lphd)	Potential Sources of Flow	Time[a] (months)	Average Flow (lphd)	Maximum Flow (lphd)	Potential Sources of Flow
N (I) - Cell 1	-	-	-	Top liner leakage and construction water draining from granular collection trenches of Cells 2 through 4	12	170	190	Top liner leakage for Cells 1 through 4
					24	140	230	
					36	190	260	
Cell 2	1	470	560		6	130	150	
					12	270	470	
					23	22	25	
					33	17	20	
Cell 3	2	130	180		6	130	230	
					12	170	190	
					24	33	60	
Cell 4	2	5	10		6	6	16	
					8 to 24	0	0	
O (II) - Cell 1	-	-	-	Construction water draining from sand of Cell 3	12	5	50	Top liner leakage for Cells 1 through 3
					32	7	15	
Cell 2	-	-	-		10	30	110	
					30	2	11	
Cell 3	1	200	(c)		3	20	110	
					24	8	56	
P (II) - Cell 1	-	-	-		16	350	910	Top liner leakage for Cells 1 and 2
					22	280	500	
Cell 2	-	-	-		16	160	470	
					22	83	290	
Q (II) - Cell 1	1	310	(b)	Construction water draining from sand for Cells 1 and 2	12	260	(b)	Top liner leakage for Cells 1 and 2
					24	470	(b)	
					30	1,200	(b)	
					38(f)	110(f)	(b)	
					44(f)	120(f)	(b)	
Cell 2	1	370	(b)		12	1,100	(b)	
					24	770	(b)	
					30	1,200	(b)	
					38(f)	100(f)	(b)	
					44(f)	120(f)	(b)	
R (IV)	0.5	23,000	(b)	Construction water draining from sand and water from consolidation of clay component of top liner	12	180	(b)	Water from consolidation of clay component of top liner and top liner leakage
					18	160	(b)	
					24	80	(b)	
					27	20	(b)	

Table A-3. Measured flow rates from the leakage detection layers of the double-lined landfills. - (continued)

Landfill (Group)	QUANTITY OF LIQUIDS MEASURED							
	END OF CONSTRUCTION				DURING ACTIVE LIFE			
	Time[a] (months)	Average Flow (lphd)	Maximum Flow (lphd)	Potential Sources of Flow	Time[a] (months)	Average Flow (lphd)	Maximum Flow (lphd)	Potential Sources of Flow
S (III) - Cell 1	-	-	-	-	39(f)	200(f)	570(f)	For Cells 1 to 4, secondary compression of clay component of top liner, infiltration water from consolidation of saturated clay foundation, or top liner leakage; for Cell 5, water from consolidation of clay component of top liner
Cell 2	-	-	-		39(f)	230(f)	520(f)	
Cell 3	-	-	-		34(f)	300(f)	1370(f)	
Cell 4	-	-	-		34(f)	200(f)	460(f)	
Cell 5	-	-	-		15	260	(c)	
T (III) - Cell 1	-	-	-	-	50(f)	30(f)	210(f)	For Cells 1 and 2, water from secondary compression of clay component of top liner and/or top liner leakage; for Cells 3 and 4, water from consolidation of clay component of top liner
Cell 2	-	-	-		48(f)	40(f)	180(f)	
Cell 3	-	-	-		1	90	200	
Cell 4	-	-	-		5	270	840	
					1	20	40	
					5	60	230	
U (III)	0.2	710	(c)	Construction water draining from granular collection trenches	6	330	470	Water from consolidation of clay component of top liner
					12	28	180	
					18	100	310	
V (II) - Cell 1	6(g)	1390	4730	For Cells 1 and 2, construction water draining from sand; top liner leakage is known to be occurring for Cell 1, which had a localized lining system failure on the landfill side slope shortly after the end of construction	-	-	-	-
Cell 2	6(g)	250	920		-	-	-	
W (II)	6(g)	16	135	Construction water draining from sand	-	-	-	-

Notes:

(a) Time interval between end of construction and flow measurements.
(b) Data is limited to total flow quantities measured over a one-month period.
(c) Data is too limited to evaluate.
(d) Waste not yet placed in cell.
(e) Cell 1 was covered with an HDPE geomembrane at 19 months after construction.
(f) Cell was closed with a cover system.
(g) The data is reported as end of construction because the hydraulic conductivities of the leakage detection layers of these facilities are so low (i.e., 10^{-5} m/s) that significant construction water drainage may still be occurring six months after construction.

Table A-4. General information on double-lined surface impoundments.

Surface Impoundment (Group)	Location (U.S.)	Average Annual Rainfall (mm)	Ground Water Separation Distance (m)	Area of Cell (hectares)	Maximum Liquid Height (m)	Drain Collector Spacing (m)	Base Slope Gradient	End of Construction	Third-Party CQA of Gmb Installation	Method of Flow Rate Measurement
A (III)	Unknown	Unknown	120	0.5	Unknown	Unknown	Unknown	6/1986	Unknown	Daily pumping if liquid present in sump, liquid volume recorded from accumulating flow meter
B (IV) - Pond 1 Pond 2	South West	280	90	0.5 0.4	3 2	25 55	0.02 0.02	6/1984 12/1984	Yes	Daily pumping with jet pump if liquid depth in sump > 0.3 m, liquid volume recorded from accumulating flow meter
C (III) - Pond 1 Pond 2 Pond 3	South West	280	90	0.5 0.7 0.3	4 3 4	25 30 20	0.02 0.02 0.02	6/1985 12/1985 6/1986	Yes	Daily pumping with jet pump if liquid depth in sump > 0.3 m, liquid volume recorded from accumulating flow meter
D (II)	East Central	1,120	Unknown	0.2	6	10	0.02	1985	Yes	Unknown
E (I) - Pond 1 Pond 2 Pond 3 Pond 4 Pond 5	South Central	480	Deep	0.4 0.6 0.4 1.1 1.1	4 4 4 Unknown Unknown	45 45 45 Unknown Unknown	0.02 0.02 0.02 0.02 0.02	11/1988 12/1988 1/1989 4/1990 6/1990	Yes	Daily pumping if liquid present in sump, liquid volume measured indirectly by multiplying pumping time by pump capacity
F (I)	Gulf Coast	1,550	30	0.8	5	10	0.003	12/1983	Unknown	Daily observation of flow from leakage detection layer collector pipe to sump
G (II)	North Central	880	Unknown	1.9	5	30	0.01	1/1989	Yes	Liquid volume removed from sump by multiplying the number of times the pump automatically activated by the volume of liquid that is stored in the sump between the "pump on" and "pump off" levels

Table A-5. Properties and thickness of the components of the double-lined surface impoundments[a].

SURFACE IMPOUNDMENT (GROUP) AND LINING SYSTEM DESCRIPTION	TOP LINER[b]				LEAKAGE DETECTION LAYER[b] (side slope)			LEAKAGE DETECTION LAYER[b] (base slope)			BOTTOM LINER[b]			
	k_{st} (m/s)	D_{st} (m)	Polymer	T_{gt} (mm)	Material	k_{dl} (m/s)	D_{dl} (m)	Material	k_{dl} (m/s)	D_{dl} (m)	k_{sb} (m/s)	D_{sb} (m)	Polymer	T_{gb} (mm)
A (III) composite/composite	10^{-10}	0.5	HDPE	1.5	Geonet	$>10^{-2}$	6×10^{-3}	Geonet	$>10^{-2}$	6×10^{-3}	10^{-9}	0.9	HDPE	1.5
B (IV) composite/composite	10^{-9}	0.5	HDPE	1.5	Sand	10^{-5}	0.5	Sand	10^{-5}	0.5	10^{-9}	0.6	HDPE	1.5
C (III) composite/composite	10^{-9}	0.5	HDPE	2.5	Geonet	$>10^{-2}$	6×10^{-3}	Gravel Overlying Geonet	5×10^{-3} $>10^{-2}$	0.3 6×10^{-3}	10^{-9}	0.9	HDPE	1.5
D (II) geomembrane/composite	-	-	HDPE	1.5	Geonet	$>10^{-2}$	6×10^{-3}	Sand	10^{-5}	0.3	10^{-9}	0.6	HDPE	1
E (I) geomembrane/composite	-	-	HDPE	2.5	Geonet	$>10^{-2}$	6×10^{-3}	Geonet	$>10^{-2}$	6×10^{-3}	10^{-9}	0.9	HDPE	1.5
F (I) geomembrane/geomembrane	-	-	CSPE	1.1	Geotextile[c]	3×10^{-3}	3×10^{-3}	Geotextile	3×10^{-3}	3×10^{-3}	10^{-2}	0.3	PVC	0.7
G (II) geomembrane/geomembrane	-	-	Ethylene Interpolymer Alloy	0.9	Sand	$>10^{-5}$	0.3	Sand	$>10^{-5}$	0.3	Needlpunched Geotextile		HDPE	1

Notes:

(a) The reported properties were obtained from engineering drawings, project specifications, and surface impoundment owners or operators. If the properties of a project specific geonet drainage layer could not be determined, the geonet was assumed to have a hydraulic conductivity greater than 10^{-2} m/s and a thickness of 6×10^{-3} m.

(b) Abbreviations are as follows: k_{st} = hydraulic conductivity of compacted clay component of top liner; D_{st} = thickness of compacted clay component of top liner; T_{gt} = thickness of geomembrane component of top liner; k_{dl} = hydraulic conductivity of leakage detection layer drainage material; D_{dl} = thickness of leakage detection layer drainage material; k_{sb} = hydraulic conductivity of compacted clay component of bottom liner; D_{sb} = thickness of compacted clay component of bottom liner; and T_{gb} = thickness of geomembrane component of bottom liner.

(c) For the groupings in the paper, the geotextile drainage layer is assumed to be equivalent to the geonet drainage layer.

Table A-6. Measured flow rates from the leakage detection layers of the double-lined surface impoundments.

Surface Impoundment (Group)	QUANTITY OF LIQUID MEASURED							
	END OF CONSTRUCTION				DURING ACTIVE LIFE			
	Time[a] (months)	Average Flow (lphd)	Maximum Flow (lphd)	Potential Sources of Flow	Time[a] (months)	Average Flow (lphd)	Maximum Flow (lphd)	Potential Sources of Flow
A (III)	2	53	(b)	Water from consolidation of the clay component of the top liner	-	-	-	-
B (IV) - Pond 1	-	-	-	-	30 36 48	220 1,200 100	(c) (c) (c)	For Pond 1, water from consolidation of the clay component of the top liner and at 36 months after construction, top liner leakage, based on testing of chemical constituents in flow; flows decreased after 36 months as the liquid height in the impoundment was decreased
Pond 2	-	-	-	-	20 to 43	0	0	
C (III) - Pond 1	-	-	-	Water from consolidation of the clay component of the top liner and construction water draining from granular collection trenches	18 to 24 25 to 28 28 to 58	0 200 0	(c) (c) (c)	For Ponds 1 through 3, water from consolidation of the clay component of the top liner and top liner leakage; for Pond 2, increased flow at 42 months after construction coincided with an increased liquid height in the pond; a geomembrane top liner defect was found on the side slope of Pond 2 and repaired
Pond 2	-	-	-		18 30 42 50	230 40 120 40	(c) (c) (c) (c)	
Pond 3	2	930	-		12 24 36	160 45 14	(c) (c) (c)	
D (II)	-	-	-	-	7 >8	64 0	(b)	Top liner leakage until 7 months; no flow after 7 months (reason unknown)
E (I) - Pond 1	1	0	0	No flow recorded	19	0	0	For Pond 2, top liner leakage due to 20-mm long defect at a geomembrane seam on side slope of pond 15 months after construction; the defect was repaired
Pond 2	1	0	0		15 18	30,000 0	(c) 0	
Pond 3	1	0	0		17	0	0	
Pond 4	1	0	0		3	0	0	
Pond 5	1	0	0		3	0	0	
F (I)	0.5 2	400 48	460 300	Construction water draining from granular collection trenches and, at 0.5 months after construction, top liner leakage; a geomembrane defect was found and repaired; no top liner leakage at 2 months after construction based on testing of chemical constituents in flow	12 to 47 48 to 52 53	2 1 4	(c) (c) (c)	Construction water slowly draining through geotextile leakage detection layer from 12 to 52 months after construction; based on testing of chemical constituents in flow, leakage did not occur until 53 months after construction, when the maximum liquid height in the pond was increased
G (II)	-	-	-	-	7 10	80 27	(b) (b)	Construction water draining from sand and/or top liner leakage

Notes:

(a) Time interval between end of construction and flow measurements.
(b) Data is limited to flow quantities measured over a one-month period.
(c) Data is too limited to evaluate.

FACTORS CONTROLLING THE LONG-TERM PROPERTIES OF CLAY LINERS

James K. Mitchell[1], F.ASCE, and Makram Jaber[2], A.M.ASCE

Abstract

The properties of clay liners and their susceptibility to changes with time are major concerns in design for waste repositories. Both the desired and required properties of clay liners are reviewed in this paper, and the factors affecting their long-term stability are discussed. These factors fall into four major groups: (1) soil composition, including gradation, plasticity index, and type of mineral; (2) placement conditions and construction effects, including moisture content and density, compaction method and effort, size of clods, bonding between lifts, excessive wetting, desiccation, and frost conditions; (3) post-construction changes, including change in confining stress, clogging due to fine particle migration or biological activity, piping, differential settlements below the liner, the effects of freezing, and slope instability; (4) chemical compatibility between the clay liner and the contained waste.

Introduction

Fine-grained soils are widely used as major components of liner systems in waste containment barriers, both for municipal and hazardous waste landfill leachates and for liquids in surface impoundments. Typical cross-sections of clay liner systems are shown in Figure 1. The properties of clay liners and their susceptibility to changes with time are major concerns in design. The purpose of this paper is to review both the desirable and required (by regulation) properties of clay liners, and

---

[1]Edward G. and John R. Cahill Professor of Civil Engineering, Univ. of California, Berkeley, CA 94720.

[2]Assistant Project Engineer, GeoSyntec Consultants (formerly GeoServices Inc. Consulting Engineers), 5950 Live Oak Parkway, Suite 330, Norcross, GA 30093.

to discuss the factors controlling their long-term properties.

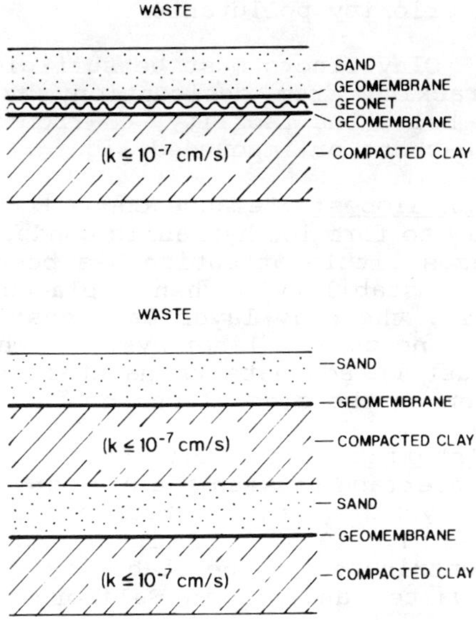

Figure 1 Typical cross-sections of liner systems that include compacted clay layers.

Desirable Properties of Clay Liners

Low Hydraulic Conductivity: Low hydraulic conductivity is a prime reason for choosing clay as a major component of waste containment barrier systems. Much care must be taken in choosing the material and in constructing the liner in a way that achieves and maintains a low hydraulic conductivity. This property may vary over several orders of magnitude for a given soil, and it can be affected greatly by details of placement conditions, construction methods, and post-construction changes in ambient conditions.

Low Diffusivity: Contaminants can migrate through clay liners by molecular diffusion driven by the chemical gradient across the liner. Molecular diffusion is independent of hydraulic gradient and can result in release of contaminants from a facility even if hydraulic flow is towards the inside of the liner (Shackleford, 1988). Molecular diffusion may be the dominant transport mechanism across a well designed and constructed clay liner, i.e., if its hydraulic conductivity is very small. For example, Johnson et al. (1989) showed that, for a typical 1 m (3 ft) thick clay liner, molecular diffusion alone can result in short (about 5 years) breakthrough times, and in steady fluxes that may be significant for the common priority pollutants.

Ductility: Clay liners must be sufficiently ductile to prevent cracking from desiccation, from differential settlement below the facility, or from interaction with landfill leachate or impounded liquid.

Stability on Slopes: Because compacted clay liners are built mainly to form low hydraulic conductivity barriers, in some cases little attention has been given to their mechanical stability when placed on slopes. Nevertheless, the clay layer may constitute one of the weakest components of a liner system. The compacted clay strength must be adequate to maintain stability of the liner system on the side slopes of the facility.

Adequate Interface Strength: The sliding resistance along the interface between the clay and other components of a liner system (e.g., geomembrane), can be very low (Mitchell et al., 1990). Low interface shear resistance may have contributed to the stability failure of a large hazardous waste landfill in California (Seed et. al., 1990).

Long-Term Stability: The long-term stability of a clay liner is of major concern in a facility that is to contain waste for several decades. Long-term mechanical instability of a clay liner due to creep in the clay or along the interface between the clay and other components of the liner system is a possibility that should be considered. However, the major long-term concern for clay liners is their chemical "compatibility" with leachates or impounded liquids rich in organic and inorganic compounds. It is important that these compounds not interact with the clay in a way that adversely affects its performance.

Chemical Retardation: Chemical retardation processes include adsorption, precipitation, biological processes, oxidation-reduction reactions, acid-base reactions, and complex formation and chelation (Dunn, 1983). Some of the processes (e.g., adsorption) will only retard pollutant release to the environment until an equilibrium is reached in the soil. Other processes, such as precipitation and biological clogging may lead to a decreased pore size within the clay, and therefore to a decrease in hydraulic conductivity with time.

High Effective Porosity: Olsen (1962) has shown that macropores and preferential flow paths within a compacted clay account for most of the flow through the clay. "Effective porosity" is defined as the pore volume that is effective in conducting flow divided by the total volume of a clay sample. The effective porosity of a compacted clay is smaller than, or equal to, its total porosity. The seepage velocity of a fluid through a porous medium in the direction of the hydraulic gradient is inversely proportional to the effective porosity, i.e.:

$$v_s = \frac{v}{n_e} = \frac{ki}{n_e}$$

where v_s is the seepage velocity, v is the average Darcy velocity, n_e is the effective porosity, k is the hydraulic conductivity, and i is the hydraulic gradient. The breakthrough time, t_b, (given by $t_b = d/v_s$, where d is the thickness of the liner, and diffusion and retardation effects are neglected) of pollutants through a clay liner is, therefore, inversely proportional to effective porosity. It is desirable to compact the clay in a way to obtain an effective porosity that approaches its total porosity, thereby maximizing the breakthrough time. Recent research by Daniel et al. (1990) has shown that the effective porosity is a function of hydraulic gradient, with lower values of n_e associated with lower values of gradient.

Constructibility: A clay liner should be constructed using a soil that can be processed and compacted to the specified uniformity, dry density, and moisture content without excessive difficulty. Adequate trafficability on top of both the uncompacted and compacted material should also be considered.

Required Properties of Clay Liners

U.S. Environmental Protection Agency guidelines (EPA, 1985) require that the compacted clay component of a double liner system be at least 1 m (3 ft) thick, with a maximum saturated hydraulic conductivity of 1×10^{-7} cm/s.

A recent EPA technical resource document (TRD) (Goldman et al. 1988) contains a thorough discussion of clay behavior and its use in compacted clay liners. While the basic regulatory requirement of 1×10^{-7} cm/s maximum saturated hydraulic conductivity remains, the recommendations for selecting clay materials for liners were expanded, and it is suggested in the TRD that specifications for clay liners address the following items to ensure that the required hydraulic conductivity and system stability are achieved and maintained:

- Facility configuration and size.
- Foundation preparation (e.g., pulverize and recompact subgrade; smooth subgrade surface).
- Liner material characteristics (e.g., gradation, plasticity index).
- Liner thickness and permeability (e.g., thickness no less than 1 m (3 ft); hydraulic conductivity no greater than 1 x 10^{-7} cm/s).
- Side wall slope (typical side wall slope are not steeper than 3 horizontal to one vertical).
- Bottom slope and configuration (bottom slope greater than 2% to ensure drainage towards sump).
- Lift orientation on sideslopes (e.g., horizontal or continuous lifts).
- Lift thickness (e.g., not more than 15 cm (6 in) after compaction).
- Maximum clod size (e.g., clods with diameter larger than 25 to 50 mm (1 to 2 in.) should be mechanically pulverized prior to compaction).
- Degree of compaction (e.g., dry density larger than 95 percent of standard Proctor maximum dry density).
- Compaction moisture content (a range of moisture contents wet of optimum should be specified).
- Scarification between lifts.
- Compaction equipment (e.g., sheepsfoot roller, weight of roller, etc.) and number of passes.
- Test-fill to evaluate equipment and soil, and to determine which compaction procedures work best to achieve the required hydraulic conductivity.

It is clear from the above list that the EPA recognizes many of the factors that lead to satisfactory clay liner performance. It is, however, difficult to address all of the relevant factors through regulation. Ideally, the burden of ensuring that a clay liner for a specific landfill or impoundment will function properly falls on the design engineer, the regulator, and the contractor working as a team.

Factors Controlling Properties of Clay Liners

A large number of factors may influence the long-term properties of clay liners. In this paper, these factors are organized into four major groups: (i) composition; (ii) placement conditions and construction effects; (iii) post-construction changes; and (iv) chemical compatibility.

Composition: The composition of a soil determines the possible range of its properties. In order to obtain a clay liner that will perform satisfactorily, i.e., be stable and have a hydraulic conductivity smaller than 1 x 10^{-7} cm/s, the material used to construct the liner must be chosen carefully. Soils that are appropriate for use in liner construction should have at least 20% fines, no more than 10% gravel size, and no soil particles or chunks of diameter larger than 25 to 50 mm (1 to 2 in.) (Goldman et al., 1988). Well graded soils are easiest to compact to high densities with a uniform distribution of small pores. The plasticity index (PI) should be larger than 10 percent. However, if the PI is larger than 30 to 40 percent, the material becomes sticky when wet and may form large clods. It therefore becomes difficult to construct, as experienced both in the laboratory and in the field by Elsbury et al. (1988).

The soil mineral types determine potential ranges of soil properties, including plasticity, shrink-swell, adsorption, ion exchange, flocculation-deflocculation, etc. Soil minerals of all types and their physical and chemical properties are described in detail in Dixon and Weed (1989). The most important minerals in the soil relative to the attainment and maintenance of low hydraulic conductivity, durability, and long-term stability are the clay minerals. Smectite (montmorillonite) might at first seem best. However, the very high plasticity and water holding capacity of expanded smectites makes them susceptible to collapse when exposed to chemicals of various types. This collapse could lead to increase in hydraulic conductivity. Thus, non-expansive illitic and kaolinitic clays may be preferable in many cases.

Soil minerals that may be soluble in the chemicals from a waste containment facility should not be used in liner construction. The authors are aware of a case in Kansas, in the early 1980's, where a 150 mm (6 in.) thick liner for a liquid impoundment was constructed using native carbonate soil over a limestone subgrade. Chromic acid was disposed of in the impoundment. The carbonate soil was dissolved by the acid, and large quantities of industrial waste were released to the environment. Compatibility between clay and chemical waste is discussed in more detail later.

Placement Conditions and Construction Effects: The placement conditions and the details of construction of a clay liner determine the fabric and structure of the compacted clay and, therefore, its actual properties, including hydraulic conductivity and strength. As noted by Mitchell and Madsen (1987), three levels of fabric may be important when dealing with fine-grained soils as barriers to the flow of chemicals: (i) the microfabric, consisting of the regular aggregations of particles and the very small pores within the aggregates through which little fluid will flow; (ii) the minifabric, containing these aggregations and the larger inter-assemblage pores through which flows will be much greater than through the intra-aggregate pores, since the hydraulic conductivity varies with the square of the pore radius; and (iii) a macrofabric, containing cracks, fissures, root holes, laminations, etc., through which the flow rate is so great in comparison with the other flows as to largely obscure them. Placement conditions and construction procedures must be specified and implemented in a way that eliminates the macrofabric and minimizes the effects of the minifabric. A high density liner with uniform small pores composed of a non-shrinking soil makes it difficult for particles to move or cracks to form. Such a liner is expected to be stable in the long-term, unless solutioning occurs.

The factors that control the fabric of a compacted clay and, therefore, its properties are:

- Moisture content and density: As shown in Figure 2, differences in moisture content and dry density of a compacted clay can result in differences of several orders of magnitude in hydraulic conductivity. Placement conditions that result in a high dry density and a moisture content wet of the line of optimums lead to the lowest values of hydraulic conductivity.

- Compaction method and effort: Different methods of compaction can lead to large differences in hydraulic conductivities for clay compacted wet of optimum water content to the same dry density Mitchell et al. (1965). For example, samples prepared by kneading compaction, a method that results in a more dispersed structure due to the introduction of large shear strains during compaction, have lower hydraulic conductivity than samples prepared by static compaction. For this reason, a heavy sheeps-foot roller that introduces large shear strains during compaction in the field should be used. Furthermore, the liner should be compacted in thin layers so that the feet of the compactor can penetrate and impart shear strains to the full layer.

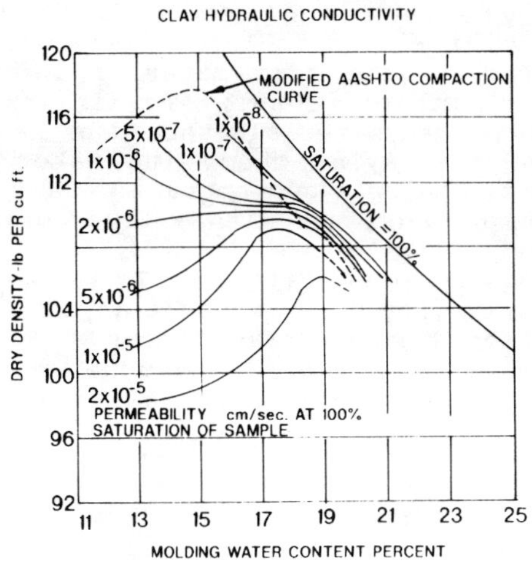

Figure 2 Contours of equal hydraulic conductivity for samples of silty clay prepared by kneading compaction (from Mitchell et al., 1965).

- Size of clods: The size of pores in the minifabric of compacted clay depends on the size of particle aggregations, or clods. The larger the clods, the larger the inter-aggregate pores, and the larger the hydraulic conductivity of the compacted clay (Olsen, 1962). The effect of clod size on hydraulic conductivity of a compacted clay is illustrated by Daniel (1984): the hydraulic conductivity of a compacted clay increased from 9 x 10^{-9} cm/s to 3 x 10^{-7} cm/s for an increase in average clod size from 1.5 mm to 9.4 mm (1/16 in. to 3/8 in). Clay clods should be destroyed in the field either before compaction, by mechanically pulverizing them, or during compaction, as suggested by Benson and Daniel (1990). Destruction of clay clods during compaction may be achieved: (i) by compacting the soil at a high molding water content so that the clods are made soft and weak, and can be easily crushed during compaction (provided the soil is reasonably workable at high water content); or (ii) by compacting the soil at a lower water content but with an extremely heavy roller that crushes the clods.

- Desiccation: Desiccation and cracking of clay liners due to drying after compaction may be a primary reason for measuring in-situ hydraulic conductivities up to 1000 times larger than values measured on the same material in the laboratory (Day and Daniel, 1985). To minimize desiccation, the compacted materials should be kept moist at all times by spraying them with water or bituminous sealers, use of a geomembrane cover, or by covering the completed liner with a layer of another soil.

- Bonding between lifts: The interface between successive lifts may present an easy path for fluid migration through clay liners. Full-scale tests (Brown et al., 1983) were done where a fluid was shown to penetrate the liner through a vertical crack, then travel along the interface between two clay lifts until it encountered another vertical crack, as shown in Figure 3. Such defects in the liner provided the fluid with a series of preferential paths that resulted in a much shorter breakthrough time and larger hydraulic conductivity than expected. Scarification and water content control at the surface of the clay after compaction of each lift can minimize horizontal preferential flow paths.

- Excessive wetting: While compacted clay liners should be kept moist after construction in order to prevent desiccation, excessive wetting of the liner may have adverse effects on the strength of the material and on the sliding resistance along the interface between the clay and the adjacent layer of the liner system. This could compromise the stability of the liner system at a later time.

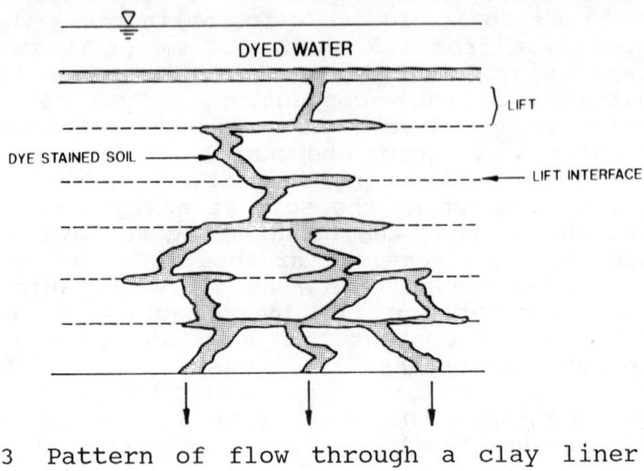

Figure 3 Pattern of flow through a clay liner (from Brown et al., 1983).

Due to the large number of construction variables determining the properties of clay liners, a field test section should be constructed prior to construction of the actual liner. The test pad is used to confirm that the materials and construction techniques proposed by the contractor result in a clay liner that has the required properties. Typical variables addressed during test pad construction are compaction moisture content, lift thickness, number of passes of the compaction equipment, methods for breaking down clods, methods for wetting or drying the clay, and construction quality control testing procedures.

Post-Construction Changes: The factors listed in the previous section are all controllable by appropriate choice of soil type and use of proper construction procedures. A carefully designed and implemented construction quality assurance plan can result in a liner that initially satisfies or exceeds the required hydraulic conductivity of 10^{-7} cm/s (Daniel, 1987; Gordon et al., 1989; Lahti et al., 1987; and others). However, changes may occur after construction (sometimes a long time after construction) that may adversely affect the properties of compacted clay. These changes may be caused by:

- Change in confining stress: In most cases, the overburden on top of the liner increases due to the weight of waste in a landfill or liquid in a surface impoundment. The resulting increase in confining stress can consolidate the clay, heal some cracks that may have developed in the liner, and, therefore, decrease its hydraulic conductivity (Fernandez and Quigley, 1990). It is important to note, however, that a decrease in effective confining stress could occur due to an increase in the level of leachate above the liner if the leachate collection system fails.

- Particle Migration: The seepage forces resulting from high gradients on top of a liner may result in particle movement in the compacted clay. Mitchell and Younger (1967) observed a drop in hydraulic conductivity of compacted clay under high gradients due to particle migration and clogging of the pores. However, if the clay is erodible, the displaced particles may be washed out and the liner may fail by piping.

- Differential settlement: Differential settlement under a clay liner as a consequence of landfill loading can result in cracking of the clay layer.

This effect is minimized if the clay liner is thick and ductile.

- Freezing conditions: Repeated freezing and thawing of a clay liner can result in a significant increase in hydraulic conductivity of the clay and a loss of strength. In tests on four fine-grained soils, Chamberlain and Gow (1979) found that repeated freeze-thaw cycles caused significant structural changes in the clay, such as severe cracking, that resulted in up to two orders of magnitude increase in their hydraulic conductivities. Similar results were presented more recently by Zimmie and La Plante (1990) as shown in Figure 4. Accordingly, clay liners should always be protected from freezing. This is most commonly done by placing a layer of cover soil and/or waste over them that has a thickness greater than the local depth of frost penetration.

- Biological clogging: The growth of micro-organisms in the slow-moving fluids through a clay liner can result in pore clogging, and therefore in a decreased hydraulic conductivity (e.g., Quigley et al., 1987).

- Slope instability: A slope stability failure that involves breach of liner can result in the localized release of contaminants to the environment. Slope instability or excessive downslope movements can result in cracking of a clay liner. Such instability can develop as waste is added to the landfill, or can be triggered by seismic activity. Slope stability must be ensured by use of appropriate side slope angles, liner system materials, and waste placement techniques to prevent interface slip within the liner system. High interface friction geomembranes, such as embossed or textured geomembranes, are now available. Tensile reinforcements, such as geogrids anchored at the top of the slope, have also been used to increase the stability of landfill liners (Giroud and Beech, 1989).

- Thixotropic changes: Mitchell et al. (1965) and Dunn (1983) observed up to an order of magnitude increase in hydraulic conductivity of compacted clay during the first month or two after compaction. They attributed this increase to thixotropic changes in the soil fabric, resulting from repulsion between the negatively charged clay particles and their rearrangement to a more

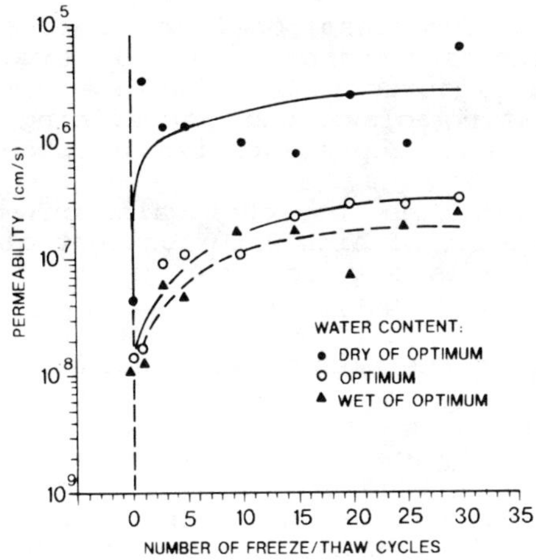

Figure 4. Summary of the 1D Freeze/Thaw Cycle Effects on Niagara Clay (from Zimmie and La Plante, 1990).

flocculated fabric. Boynton and Daniel (1985) found this effect to be much smaller for the clay used in their tests and concluded that thixotropic changes in the hydraulic conductivity of compacted clays may be material-specific. In all probability, this effect will be small in clays compacted to high density with uniform small pores because of the physical constraints to particle movements.

Chemical Compatibility: Concern over the potential detrimental effects of exposure of clay liners to liquid chemical wastes and landfill leachates has spawned a large number of studies of compacted clay - chemical compatibility. Mitchell and Madsen (1987) presented a summary and review of chemical effects on clay hydraulic conductivity. The main conclusions were:

- The three types of laboratory tests that are used for hydraulic conductivity measurement each have inherent advantages and limitations (Daniel et al., 1985). The rigid wall permeameter system will overestimate hydraulic conductivity whenever chemical-clay interactions cause shrinkage and cracking; however, it is well-suited for qualitative determinations of whether or not there may be adverse interactions. On the other hand, the flexible wall system can lead to an underestimate of the hydraulic conductivity since the lateral confining pressure prevents cracks from opening. The consolidometer permeameter system may offer the best means for quantitative assessment of chemical interactions, induced volume changes, and for testing clays under confining stress-states that are most representative of those in the field.

- The influences of chemicals on the hydraulic conductivity of high water content clays such as in slurry walls may be much greater than on lower water content compacted clays owing to the greater particle mobility and easier opportunity for fabric changes in the high water content system.

- The effects of inorganic chemicals on hydraulic conductivity are consistent with: (1) their effects on the double layer and interparticle forces that promote flocculation, dispersion, shrinkage and swelling; (2) their effects on surface and edge charges on particles and the influences of these charges on flocculation and deflocculation; and (3) their effects on pH. Acids promote solutioning of carbonates, iron oxides, and the alumina octahedral layers of clay minerals. Bases promote solutioning of the silica tetrahedral layers and, to a lesser extent, alumina octahedral layers of the clay minerals. Removal of the dissolved material can cause increases of hydraulic conductivity; whereas, precipitation of this material can cause pore clogging and hydraulic conductivity decreases.

- The most important factors controlling the influences of organic chemicals on hydraulic conductivity are: (1) their water solubility; (2) their dielectric constant; (3) their polarity; and (4) whether the clay is exposed to the pure organic or a dilute solution. Many pure organic liquids interact adversely with clays by causing some shrinkage and cracking, with concurrent large increases in hydraulic conductivity. However, if the compacted clay is densely compacted and does not contain expanded clay minerals, and is maintained under a confining stress, the increase may not be significant. In practice, exposure of clay barriers to water-insoluble pure or concentrated organics is likely only in the case of spills, leaking tanks, and with organic dense non-aqueous phase liquids (DNAPLs) or "sinkers" that accumulate at low spots above liners.

Contrary to the inorganic compounds, dilute solutions of organic compounds have essentially no effect on the hydraulic conductivity of clays.

However, this does not mean that they will not be transported through liners. Unless adsorbed by the clay itself or by organic matter in the clay, they will migrate with the seepage water by advective transport.

- The information so far available on the effects of organic compounds on hydraulic conductivity is comprised almost totally of test results for simple mono-chemical systems and test programs that have extended only over very short time periods relative to the active life and post-closure care periods of most waste containment and storage facilities. More research on the effects of mixed chemical systems and the effects of prolonged exposure is needed.

Research conducted since 1987 has addressed some of the issues raised by Mitchell and Madsen. Quigley et al. (1987) conducted consolidometer permeability tests on a natural clay from the 1.5 m (4.5 ft) contaminated zone below a 15 year old domestic waste landfill. The dominant chemical transport mechanism was diffusion. Their main findings were that the hydraulic conductivity of the clay was not significantly altered by interaction with dissolved chemicals, and that heavy metal precipitation and increased bacterial activity in the clay near the interface with waste has resulted in a slight decrease in measured hydraulic conductivity.

Fernandez and Quigley (1988) showed that both viscosity and dielectric constant affect the hydraulic conductivity of clayey soil permeated with water-soluble organics under no vertical stress. The viscosity of the permeant increases and its dielectric constant decreases as the percentage of organics in the permeant increases. As illustrated in Figure 5(a), the increased viscosity of the permeant, at hydrocarbon contents less than 70%, led to a decrease in hydraulic conductivity. It is interesting to note, however, that the decrease in dielectric constant of the permeant up to that point had no effect on the intrinsic permeability, as shown by Figure 5(b). The intrinsic permeability K is given by:

$$K = k\,\mu/\gamma$$

where μ and γ are the viscosity and unit weight of the permeant, respectively. Use of the intrinsic permeability normalizes the results for viscosity and density differences among permeants.

That the hydrocarbon had no effect at concentrations

smaller than 70% is attributed to the greater affinity of the clay particles to the more polar water molecules, and the resulting exclusion of organic molecules from the double layer. The effect of decreased dielectric constant becomes large for a "hydrocarbon" content larger than 70% and results in about three orders of magnitude increase in hydraulic conductivity under zero vertical stress.

Fernandez and Quigley (1990) showed, however, that this increase in hydraulic conductivity can be prevented by application of a confining effective stress either before or after permeation of a compacted clay by concentrated hydrocarbons (Figure 6). Differences in the hydraulic conductivity values between water and the leachate-ethanol mixtures could be ascribed to differences in viscosity. A larger effective stress was needed to "heal" the clay after it was permeated at zero confining stress than to prevent chemical cracking altogether by applying the effective stress before permeation.

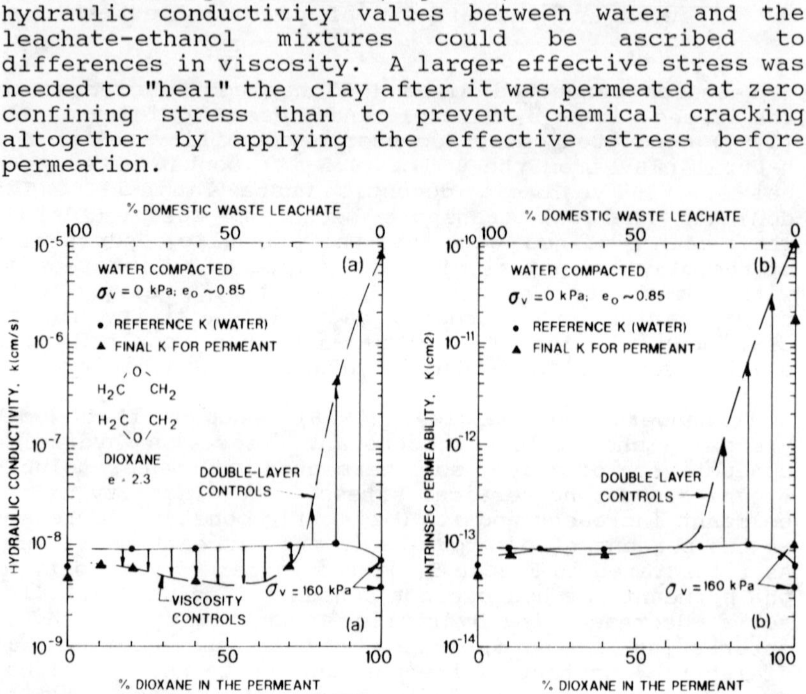

Figure 5 Hydraulic Conductivity (a) and intrinsic permeability (b) of water-compacted brown Sarnia Clay permeated with leachate-dioxane mixtures (from Fernandez and Quigley, 1988).

The performance of 15 landfills and impoundments, mostly containing hazardous waste, was reviewed by Goldmann et al., (1988). It was stated that most performed satisfactorily. However, in one case, groundwater contamination resulted, possibly from reaction between the low pH leachate and calcium carbonate inclusions in the local soil. In another case, the presence of heavy chlorinated organic liquids that concentrated on top of a clay liner resulted in its failure.

"Possible effects from organic solvents" were believed to be the cause for a measured hydraulic conductivity of a clay liner larger than 1 x 10^{-7} cm/s (Daniel, 1987). Ten other cases of compacted clay liner failure were reported by Daniel (1987), but they were attributed to poor construction of the liner.

Herzog et al. (1989) reported that reactions of organic chemicals may have been a contributing factor to ground-water contamination from the Wilsonville landfill in Illinois.

Too few research results and case histories on chemical compatibility of compacted clays have been reported since 1987 to confirm the conclusions reached by Mitchell and Madsen (1987) that serious problems are

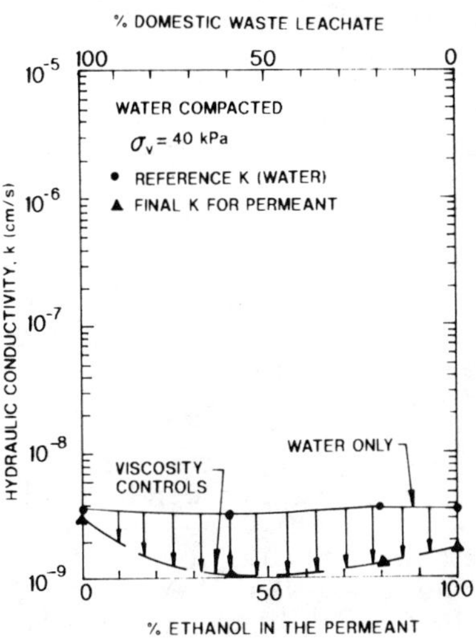

Figure 6 Hydraulic Conductivity of water-compacted clay permeated with leachate-ethanol mixtures. The clay was prestressed at a vertical effective stress of 40 kPa (from Fernandez and Quigley, 1990).

likely only when concentrated hydrocarbons come in contact with a clay liner under no, or very small, overburden, or if the soil minerals are soluble in the type of leachate generated. However, the authors are unaware of any findings that serve to refute them.

The compatibility of a clay liner material with leachate can be assessed in the laboratory, using consolidometer tests under the overburden stresses expected in the field, and preferably using a permeant that is as close in composition as possible to the leachate or liquid waste expected in a specific waste containment facility. Several pore volumes of leachate should flow through the test samples to simulate a long time of permeation in the field.

Performance Monitoring

Whether the properties of clay liners that are properly constructed are maintained over the long-term can only be proven in the field through monitoring of the liner performance. Collection basin lysimeters should be installed for that purpose at all sites using clay liners, as has been the case in the state of Wisconsin since 1980 (Gordon et al., 1989). The hydraulic conductivities of three relatively thick clay liners (1.2 to 1.5 m) (4 to 5 ft)) reported by Gordon et al. (1989) were all in the range of 1×10^{-8} cm/s to 1×10^{-9} cm/s and were not affected by the leachate over a period of 5 to 7 years.

Basin lysimeters are part of a long-term monitoring plan of the Keele Valley Landfill at Maple, Ontario (Lahti et al., 1987). The average measured hydraulic conductivity of the 1.2 m (4 ft) thick liner from 1983 to 1985 was about 1×10^{-8} cm/s, and it remained essentially constant during this period.

The Illinois Geologic Survey is currently conducting a long-term field hydraulic conductivity test on a 15 m (45 ft) long, 8 m (24 ft) wide, and 1 m (3 ft) thick clay liner submerged beneath 0.3 m (1 ft) of water (Krapac et al., 1990). The test was started in April 1988 and no breakthrough has yet been observed. Based on the instruments within the liner, an estimated saturated hydraulic conductivity of 6.7×10^{-8} cm/s to 2.5×10^{-9} cm/s was obtained.

As more data from these and other long-term field tests and monitoring programs become available, further understanding and confidence in the stability of clay liners over long periods of time can be gained.

Conclusions

Compacted clay liners have been and will continue to be essential components of most waste landfills and surface waste impoundments. The role of a clay liner is to contain liquid waste and leachate to the maximum possible extent. Regulatory requirements established an upper limit for the value of hydraulic conductivity and minimum requirements for thickness. However, it is the responsibility of the design engineer to see not only that these criteria are sufficient to ensure satisfactory performance, but also that the liner will continue to perform in the manner specified for many years into the future. Accordingly, the ability of the clay liner to retain its integrity and retention capability over very long time is essential.

By their very nature most clay soils are quite stable materials in their natural state, because they are towards the end point of the degradation phase of the weathering and rock-forming cycle. Thus, if a naturally occurring clay soil is compacted to high density, thereby producing a material with very low hydraulic conductivity, and if it is maintained within the same ranges of temperature, pressure, and chemical and biological environment, it would be expected to function well as a seepage barrier indefinitely. In waste containment applications, however, conditions do not remain the same. The permeation of a compacted clay liner by chemicals of many types is inevitable, since no compacted clay or any other type of liner material is either totally impervious or immune to chemical interactions of various types. In addition, most clay liner systems are subjected to distortional stresses that may cause differential movement. If these movements lead to formation of open cracks, then the liquid retention ability of the system will be lost.

The essential factors to be considered in the design and construction of a good clay liner have been reviewed in this paper. They include (1) selection of a soil that has a stable composition, is well-graded, and is of medium plasticity, (2) compaction to a high density at a water content wet of optimum using a high compactive effort that induces large shear strains, (3) anticipation of potential changes in ambient conditions over time and their possible effects on the liner, and (4) a thorough consideration of the compatibility between the waste liquids and leachates and the clay. Dealing with factors (3) and (4) is clearly the most difficult and most challenging part of the problem.

Although a general understanding of the effects of chemicals on the hydraulic conductivity of compacted clay now exists, there remains a paucity of information on the effects of mixed chemicals. The number of well-documented field cases that provide information about the long-term performance of clay liners is small. Clay liner systems have been in use for many years; however, monitoring of them has not been extensive, and only for the last ten years or so has there been intense interest in the details of liner performance. Fortunately, there appear to be no data so far to indicate that well-designed and constructed clay liners will not continue to function as intended. Continued observation and monitoring of the performance of recently constructed clay liners is essential.

Acknowledgements

The authors wish to thank Dr. Rudolph Bonaparte for his review of the paper.

References

Benson, C.H. and Daniel, D.E., (1990), "Influence of Clods on Hydraulic Conductivity of Compacted Clay", Journal of Geotechnical Engineering, Vol. 116, No. 8, pp. 1231-1243.

Boynton, S.S. and Daniel, D.E., (1985), "Hydraulic Conductivity Tests on Compacted Clay", Journal of Geotechnical Engineering, Vol. 111, No. 4, pp. 465-478.

Brown, K.W., Green, J.W., and Thomas, J.C., (1983), "The Influence of Selected Organic Liquids on the Permeability of Clay Liners", Proceedings, Ninth Annual Research Symposium on Land Disposal of Hazardous Wastes, U.S. EPA, Cincinnati, Ohio, EPA-600/9-83-018, pp. 114-125.

Chamberlain, E.J. and Gow, A.J., (1979), "Effect of Freezing and Thawing on the Permeability and Structure of Soils", Engineering Geology, Vol. 13, pp. 73-92.

Daniel, D.E., (1984), "Predicting Hydraulic Conductivity of Clay Liners", Journal of Geotechnical Engineering, Vol. 110, No. 4, pp. 465-478.

Daniel, D.E., Anderson, D.C., and Boynton, S.S., (1985), "Fixed-Wall Versus Flexible Wall Permeameters", Hydraulic Barriers in Soil and Rock, ASTM STP 874, pp. 107-126.

Daniel, D.E., (1987), "Clay Liners for Land Disposal Facilities", Geotechnical Practice for Waste Disposal '87, R.D. Woods (ed.), ASCE, New York, pp. 21-39.

Daniel, D.E., Liao, W.P., and Shackelford, C.D., (1990), "Rate of Flow of Leachate Through Clay Soil Liners", In Press, Report to the U.S. Environmental Protection Agency.

Day, S.R. and Daniel, D.E., (1985), "Hydraulic Conductivity of Two Prototype Clay Liners", Journal of Geotechnical Engineering, Vol. 111, No. 8, pp. 957-970.

Dixon, J.B., and Weed, S.B., (eds.), (1989), "Minerals in Soil Environments", Second Edition, Soil Science Society of America, Madison, WI, 1244 pp.

Dunn, R.J., (1983), "Hydraulic Conductivity of Soils in Relation to the Subsurface Movement of Hazardous Wastes", Ph.D. Dissertation, the University of California, Berkeley, CA, 338 pp.

Elsbury, B.R., Sraders, G.A., Anderson, D.C., Rehage, J.A., Sai, J.O., and Daniel, D.E., (1988), "Field and Laboratory Testing of a Compacted Soil Liner", U.S. EPA, Cincinnati, Ohio, EPA/600/2-88/067.

Fernandez, F. and Quigley, R.M., (1988), "Viscosity and Dielectric Constant Controls on the Hydraulic Conductivity of Clayey Soils Permeated with Water-Soluble Organics", Canadian Geotechnical Journal, 25(3), pp. 582-589.

Fernandez, F. and Quigley, R.M., (1990), "Controlling the Destructive Effects of Clay-Organic Liquid Interactions by Application of Effective Stresses", Research Report No. GEOT-5-90, Faculty of Engineering Science, the University of Western Ontario, London, Canada.

Giroud, J.P. and Beech, J.F., (1989), "Stability of Soil Layers on Geosynthetic Lining Systems", Proceedings, Geosynthetics '89 Conference, San Diego, U.S.A., pp. 35-46.

Goldman, L.J., Greenfield, L.I., Damle, A.S., Kingsbury, G.L., Northeim, C.M., and Truesdale, R.S., (1988), "Design, Construction, and Evaluation of Clay Liners for Waste Management Facilities", U.S. Environmental Protection Agency, Washington, D.C., EPA/530/SW-86/007F.

Gordon, M.E., Huebner, P.M., and Miazga, T.J., (1989), "Hydraulic Conductivity of Three Landfill Clay Liners", Journal of Geotechnical Engineering, Vol. 115, No. 8, pp. 1148-1160.

Herzog, B.L, Griffin, R.A., Stohr, C.J., Follmer, L.R., Morse, W.J., and Su, W.J., (1989), "Investigation of Failure Mechanisms and Migration of Organic Chemicals at Wilsonville, Illinois", Ground Water Monitoring Review, Vol. IX, No. 2, pp. 82-89.

Johnson, R.J., Cherry, J.A., and Pankow, J.F., (1989), "Diffusive Containment Transport in Natural Clay: A Field Example and Implications for Clay-Lined Waste Disposal Sites", Journal of Environmental Science and Technology, Vol. 23, No.3, pp. 340-349.

Krapac, I.G., Cartwright, K., Panno, S.V., Hensel, B.R., Rehfeldt, K.H., and Herzog, B.L., (1990), "Water Movement through an Experimental Soil Liner", Proceedings, 16th Annual Hazardous Waste Research Symposium, USEPA, April 3-5.

Lahti, L.R., King, K.S., Reades, D.W., and Bacopoulos, A., (1987), "Quality Assurance Monitoring of a Large Clay Liner", Geotechnical Practice for Waste Disposal '87, R.D. Woods (ed.), ASCE, New York, pp. 640-654.

Mitchell, J.K., Hooper, D.R., and Campanella, R.G., (1965), "Permeability of Compacted Clay", Journal of the Soil Mechanics and Foundations Division, ASCE, Vol. 91, No. SM4, pp. 41-65.

Mitchell, J.K. and Younger, J.S., (1967), "Abnormalities in Hydraulic Flow Through Fine-Grained Soil", Permeability and Capillarity of Soils, American Society for Testing and Materials, Special Technical Publications No. 417, pp. 106-139.

Mitchell, J.K. and Madsen, F.T., "(1987), "Chemical Effects on Clay Hydraulic Conductivity", Geotechnical Practice for Waste Disposal '87, R.D. Woods (ed.), ASCE, New York, pp. 87-116.

Mitchell, J.K. Seed, R.B., and Seed, H.B., (1990), "Kettleman Hills Waste Landfill Slope Failure. I: Liner-System Properties", Journal of Geotechnical Engineering, ASCE, Vol. 116, No. 4, pp. 647-668.

Olsen, H.W., (1962) "Hydraulic Flow Through Saturated Clays", Clays and Clay Minerals, Pergamon Press, London, Vol 11, pp. 131-161.

Quigley, R.M., Fernandez, F., Yanful, E., Helgason, T., Margaritis, A., and Whitby, J.L., (1987), "Hydraulic Conductivity of Contaminated Natural Clay Directly Below a Domestic Landfill", Canadian Geotechnical Journal, Vol 24, pp. 377-383.

Seed, R.B., Mitchell, J.K. and Seed, H.B., (1990), "Kettleman Hills Waste Landfill Slope Failure. II: Stability Analyses", Journal of Geotechnical Engineering, ASCE, Vol. 116, No. 4, pp. 669-690.

Shackelford, C.D., (1988), "Diffusion as a Transport Process in Fine-Grained Barrier Materials", Geotechnical News, Vol. 6, No. 2, pp. 24-27.

U.S. Environmental Protection Agency, (1985), "Draft, Minimum Technology Guidance on Double Liner Systems for Landfills and Surface Impoundments", EPA/530-SW-85-014.

Zimmie, T.F. and La Plante, C., (1990), "The Effect of Freeze/Thaw Cycles on the Permeability of a Fine-Grained Soil", Mid-Atlantic Industrial Waste Conference, Philadelphia, Pennsylvania.

Long-Term Durability and Aging of Geomembranes

Robert M. Koerner[1] , Yick H. Halse[2] and Arthur E. Lord, Jr.[3]

Abstract

Perhaps the most frequently asked question regarding geomembranes (or any other type of geosynthetic material) is, "how long will they last"? The answer to this question is illusive (in spite of a relatively large data base on polymer degradation) mainly because of the buried nature of geomembranes. Soil burial greatly diminishes, and even eliminates many of the degradation processes and synergistic effects which have been most widely investigated by the polymer industry for exposed plastics. However, different degradation processes coming from chemical interactions and extremely long time frames may be involved via exposure to liquids like leachate for systems intended to last for many decades or even hundreds of years. Thus the lifetimes of buried geomembranes can be significantly different than exposed plastics, but a quantitative method to predict "how long" is still not available.

This paper describes the various degradation mechanisms of plastics on an individual basis and then addresses the various synergistic effects which may accelerate degradation. It will be seen that synergistic effects greatly complicate the situation. While accelerated test methods are attractive to assess the various phenomena, these procedures may significantly misrepresent the actual long-term performance of geomembranes. Thus the transfer of information must proceed with caution.

[1]Bowman Professor of Civil Engineering and Director, Geosynthetic Research Institute, Drexel University Philadelphia, PA, 19104.

[2]Research Assistant Professor, Geosynthetic Research Institute, Drexel University, Philadelphia, PA, 19104.

[3]Professor of Physics, Geosynthetic Research Institute, Drexel University, Philadelphia, PA, 19104.

The paper summarizes the different geomembrane degradation processes and then concludes with some suggested procedures on long-term simulation testing coupled with observations on field behavior.

Overview of Geomembranes

Beginning with swimming pool liners (Staff 1984) made from PVC sheeting in the 1930's, and extending to bituminuous panels (Geier and Morrison 1968) for canal liners, the geomembrane era began in earnest with reservoir liners made from butyl rubber (Chuck 1970) in the 1950's. These thermoset elastomers were made from synthetic rubber which was developed during World War II. To date, some manufacturers still refer to geomembranes as "pond liners". Other names have also arisen; for example, flexible membrane liners (FML's), a term used extensively by the U.S. EPA, synthetic membrane liners (SML's), membrane liners, or, simply, liners. Being a subset of the geosynthetics area, we will refer to them as geomembranes. ASTM defines geomembranes as "an essentially impermeable membrane used with foundation, soil, rock, earth or any other geotechnical engineering related material as an integral part of a man-made project, structure or system".

Throughout the 1960's and the 1970's a tremendous variety of geomembrane formulations were developed. They were so plentiful that they almost defied classification. Many were blends of polymers and even blends of polymers and nonpolymers. Copolymers were developed which gave additional options. Even the manufacturing process gave rise to many variations. The EPA Technical Guidance Document (Matrecon, Inc. 1988) and Kay's book (1988) gives some insight into the varieties of geomembranes available during this period. The major use of geomembranes, however, did not arrive until 1982 when the U.S. EPA required that a liner must "prevent" pollution migration, rather than only "minimize" pollution migration (U.S. Federal Register 1982). This legislation essentially required the use of FML's, or geomembranes, as being preferred over clay liners.

Regarding the culling out of the large variety of geomembrane types existing at that time, another EPA document, this time in the form of a test protocol, was significant. In 1984, EPA decided that a test method for chemical compatibility, or resistance, was necessary so that all EPA Regions and State Environmental Departments would be permitting candidate geomembranes against the same test standard. This test method, known as EPA Method 9090 (U.S. EPA 1984), has had the effect of greatly reducing the many possible types of geomembranes in current use, particularly those used for pollution control. The test method has a parallel document directed at assessing the test results, which is in the form of an expert computer system called FLEX (U.S. EPA 1987). In the next section, characterization of the most widely used geomembranes is presented.

Types and Properties of Commonly Used Geomembranes

With the initial caution that polymer formulation and processing is an on-going and constantly changing technology, we will attempt to categorize the geomembranes that are in common use. To be noted, however, is that our perspective is from a geosynthetic engineering design point of view. A different grouping would indeed occur from a polymer chemist's or compound formulator's perspective.

(a) Stiff (Semi-Crystalline) Thermoplastic Geomembranes - In the stiff, semi-crystalline, thermoplastic category are those geomembranes with crystallinity near, or above, 50% which results in stiffness values of greater than 1000 g-cm as per the ASTM D-1388 flexural rigidity test. This is a bending test developed for fabrics, but it can be readily used to distinguish stiff-from-flexible geomembranes.

By far the most important geomembrane in this category is high density polyethylene (HDPE). There are a number of variations within HDPE, an important one being the development of textured sheet which results in a high friction surface. Also coextrusion manufacturing can result in very intriguing composite materials with the basic sheet being HDPE.

As a polymer formulation, HDPE is almost pure polyethylene resin (about 97%), in the 0.935 to 0.937 g/cc density range. When carbon black is added for ultraviolet stability, however, the material's overall density is at, or slightly above, the 0.941 g/cc lower limit of ASTM definition of high density polyethylene. Thus it is commonly referred to as HDPE. The balance of the compound is 2.0% to 2.5% carbon black and the remaining 1.0% to 0.5% is antioxidant and processing aide. The processing aides provide viscosity control, manufacturing lubrication, and prevent adhesion between various surfaces. They have also been called viscosity depressants, slip agents and antiblocking agents. They are generally proprietary as to their exact chemical compositions. It should also be noted that the polyethylene resins themselves have certain uniquenesses, particularly the nature and extent of tie molecule bonding and branching.

(b) <u>Flexible (Low Crystallinity) Thermoplastic Geomembranes</u> - This group of geomembranes is characterized by stiffness values in the ASTM D-1388 flexural rigidity test as being significantly lower than 1000 g-cm. The main geomembranes in this category are PVC, CPE, CSPE and VLDPE. Some typical values of flexural rigidity are the following:

- 0.50 mm polyvinyl chloride (PVC) = 4 g-cm
- 0.75 mm chlorinated polyethylene (CPE) = 20 g-cm
- 0.75 mm chlorosulfonated polyethylene (CSPE) = 25 g-cm
- 0.75 mm very low density polyethylene (VLDPE) = 78 g-cm

While all of the above geomembranes have some crystallinity, it is very low in comparison to HDPE. Yet, all are indeed thermoplastic polymers and can be seamed using thermal methods. The formulations of these materials vary widely. Some typical values follow in Table 1.

Table 1 - Typical Formulations for Flexible, Thermoplastic, Geomembranes

Geo-membrane	Resin (%)	Plasticizer (%)	Carbon Black & Filler (%)	Additive*
PVC	45-50	35-40	10-15	3-5
CPE	60-75	10-15	20-30	3-5
CSPE	45-50	2-5	45-50	2-4
VLDPE**	96-98	0	2-3	1-2

*refers to antioxidant, processing aids and lubricants
**note that this formulation is typical of most polyethylenes, including HDPE

(c) <u>Reinforced, Flexible (Low Crystallinity) Thermoplastic Geomembranes</u> - The behavior of geomembranes when they are fabric reinforced (either via an internal scrim or via spread coating) is very different than the unreinforced variety of the exact same polymer. Certainly the short-term engineering design properties are greatly altered and conceivably the long-term properties may be altered as well. The geomembranes are indeed flexible, however, as the following data indicates. The stiffness values indicated are via the ASTM D-1388 test for flexural rigidity:

- 0.91 mm chlorinated polyethylene (CPE-R) - 25 g-cm
- 0.91 mm chlorosulfonated polyethylene (CSPE-R) - 30 g-cm
- 0.66 mm ethylene interpolymer alloy (EIA-R) - 60 g-cm

As with the unreinforced materials just mentioned, all can be seamed using thermal methods and are truly thermoplastic. The formulations of the polymer component are the same as given in Table 1.

Regarding the type of fabric used for the scrim reinforcement of CPE-R and CSPE-R there are many possibilities. Woven fabrics of high tenacity polyester or nylon are the most common. They are often in a pattern of 10 yarns per inch (4 yarns per centimeter) in both directions, which is called a 10 × 10 scrim. Other variations are 6 × 6 and 20 × 20 patterns, as well as unbalanced variations. For reinforced geomembranes like EIA-R, a very tightly woven fabric is used. Not specifically mentioned are spread coated fabrics where a variety of nonwoven needlepunched fabrics can be utilized.

(d) Other Geomembranes - While the focus in this paper will be on the geomembranes just reviewed, there are many other possibilities. Two complete groups have been omitted since they are not currently used to any significant degree in North America. They are the thermoset elastomers (Matrecon 1988, Kays 1988) and polymer modified bitumens (Gamski 1984), the latter group being used quite often in Europe. Other newer materials which fall within the categories just mentioned are in the development stage and undoubtedly more will appear in the future. This review, however, will be sufficient to build upon in describing the various degradation processes and synergistic effects which may occur.

Mechanisms of Degradation

This section of the paper describes various polymer degradation processes which can act within a geomembrane. Each process and its resulting implication is taken by itself as though it were acting in isolation. This, of course, is not field representative, but we feel that it is necessary to describe the isolated events before synergistic effects can be considered. Please note at the outset that our perspective is from a geosynthetic engineering design point of view. From a chemistry or polymeric science point of view there are many references treating the various subjects in a much more rigorous (and undoubtedly enlightened) manner.

(a) Ultraviolet Degradation - As shown in Figure 1, the spectrum of natural light is broken into two major regions (visible and ultraviolet) according to the wavelength of the solar radiation. It is well established in the polymer literature that certain wavelengths within the ultraviolet portion are particularly degrading to polymeric materials. Van Zaten (1986) makes mention of the following commonly used polymers and their most sensitive wavelengths, all of which are in the ultraviolet region and are noted on Figure 1.

- polyethylene = 300 nm
- polyester = 325 nm
- polypropylene = 370 nm

Furthermore, the mechanism of degradation is also well understood. The light with the most sensitive wavelengths enters into the molecular structure of the polymer liberating free radicals which cause bond scission in the primary bonding of the polymer's backbone. This mechanism, in direct proportion to the intensity, causes a reduction in mechanical properties to the eventual point where the polymer becomes brittle and cracks to unacceptable levels.

The above type of degradation is greatly reduced by the use of carbon black or chemical based light stabilizers. Carbon black is a finely dispersed powder of approximately micron size which acts as a blocking (or screening) agent to prevent the ultraviolet light from entering into the polymer structure. It also absorbs some of the energy. Its effectiveness decreases uniformly with time of exposure so that the amount and dispersion of the carbon black is important (Apse 1989). The maximum amount, however, is limited to the amount which interferes with the growth and strength of the polymer structure. Hindered amine light stabilizers (HALS) are chemicals added to the polymer compound which react with the free radicals liberated by the ultraviolet light preventing the propagation of degradation. When such additives are consumed, however, continued ultraviolet exposure will cause rapid degradation of the polymer. A combination of carbon black and chemical absorbers has been shown to be very effective in avoiding ultraviolet induced degradation of polymers (Grassie and Scott 1985).

For geomembrane applications, a soil backfill or other covering eliminates the problem of ultraviolet degradation entirely. Only exposed geomembranes are subjected to ultraviolet degradation and as little as 15 cm of soil cover is sufficient to prevent its occurrence. Obviously, this cover soil must be placed in a timely fashion which can be achieved in all except the following

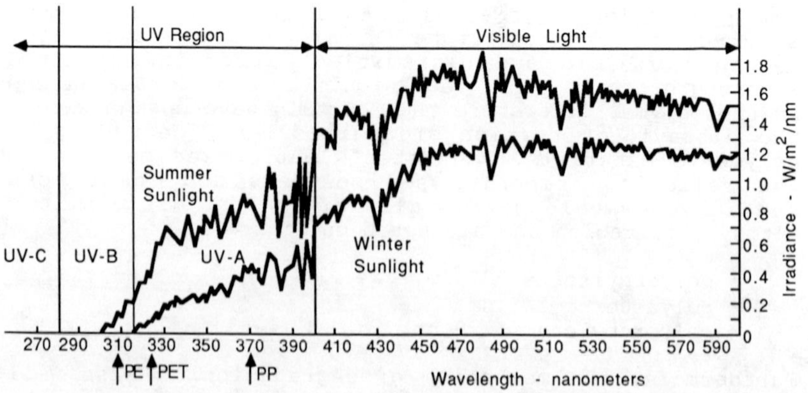

Figure 1 - The Wavelength Spectrum of Visible and UV Solar Radiation.

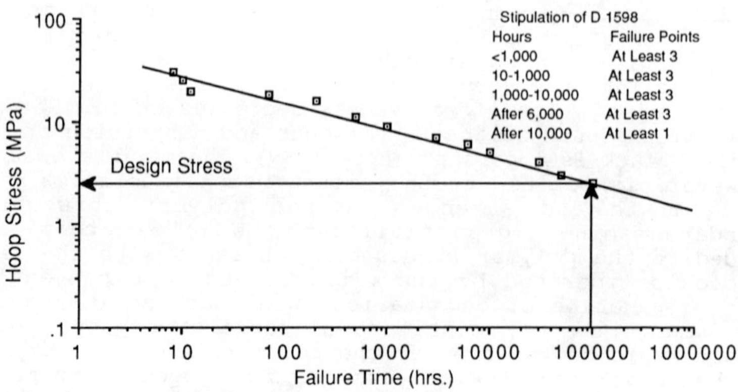

Figure 2 - Schematic Plot of Time of Failure versus Pipe Hoop Stress for Burst Testing of Un-Notched PE Pipe.

situations.

- Surface impoundments above the liquid level and along their horizontal runout length
- Canal liners above the liquid level and along their horizontal runout length
- Covers of surface impoundments, i.e., floating covers
- Landfill liners on side slopes which have had their surfaces exposed by erosion of cover soil and are inaccessible.

For the above situations of exposed geomembranes some amount of degradation over time is unavoidable.

(b) Radiation Degradation - There are a number of reviews on the effects of radiation on polymer properties (Phillips 1988, Charlesby 1960). An extremely brief summary will be given here. The effects of γ-rays, neutrons and β-rays are essentially equivalent when their different penetrating powers are considered. β-rays (electrons) penetrate about a millimeter into a polymer, whereas γ-rays and neutrons penetrate much further. α-rays (helium nuclei) penetrate only micrometers hence are only involved with very near surface damage.

The basic mechanical short term properties of a typical polymer start to change at a total radiation dose of between 10^6 and 10^7 rads (Phillips 1988). A rad is equivalent to 100 ergs of absorbed energy per gram of material. For reference purposes, the lethal dose of radiation to a human is about 100 to 200 rads. Therefore it would appear that if a geomembrane is containing low level nuclear waste of even lower radiation than the lethal human dose, the time before significant damage occurs to its short term mechanical properties will be quite long indeed. Other, more subtle changes may occur. For example, even very small amounts of local surface damage in a semi-crystalline geomembrane might cause a serious reduction in the stress crack resistance of the material. The effects of radiation on the additives may be a more severe problem than the effect on the polymer itself. It is possible that after a certain irradiation the material will be more susceptible to other degradation processes.

While no test protocol exists for evaluation, some form of incubation test method can be used with suitable modifications. Whyatt and Fansworth (1990) have evaluated a number of different geomembranes in simulated short-term tests in a high pH (≃14 weight percent NaOH) inorganic solution at 90°C and subjected them to radiation doses up to 39×10^6 rad. It was found that only polyethylene and polypropylene geomembranes were uneffected by the radiation. Furthermore, the radiation did not have a significant effect on other chemical degradation rates.

With the absence of radioactive materials in the waste material, however, the subject becomes a moot point. Fortunately, this is the case for most solid and liquid waste contained by geomembranes and related geosynthetic materials.

(c) Chemical Degradation - The reaction of various geomembranes to chemicals has probably been studied more than any other liner degradation mechanism. Most of the work is laboratory oriented via simple immersion tests but the body of knowledge is so great that a reasonable confidence level can be associated with manufacturers listings and recommendations. Complex waste streams like leachate, however, are usually not addressed and must be evaluated on a site specific basis. For this reason the U.S. EPA developed the Method 9090 procedure. Here samples of the candidate geomembrane are exposed at 23°C and at 50°C and removed at 30, 60, 90 and 120 days. Various physical and mechanical tests are performed and then compared to the unexposed geomembrane. A percent change in this behavior is calculated. When plotted for the various exposure times, trends can be established and a decision made as to the nature and degree of chemical resistance.

Depending on the type of leachate vis-a-vis the polymeric compound from which the geomembrane is made, a number of reactions may occur.

- No reaction may occur, which indicates that the geomembrane is resistant to the leachate; at least for the time periods and temperatures evaluated.
- Swelling of the geomembrane may occur which in itself may not be significant. Many polymers can accommodate liquid in their amorphous regions without a sacrifice of physical or mechanical properties. Swelling, however, is often the first stage of subsequent degradation and a small loss in modulus and strength may occur. The effect is often reversible when the liquid is removed.
- Change of physical and mechanical properties, of course, signifies some type of chemical reaction. The variations are enormous. Quite often the elongation at break in a tensile test will be the first property to show signs of change. It will first occur with the 50°C incubation data, since this can be considered to be an accelerated test over the 23°C incubation data.
- A large change of physical and mechanical properties signifies an unacceptable performance of the

geomembrane. Limits of acceptability are, however, very subjective. Table 2 gives a number of recommendations as accumulated by Koerner, 1990. It should be noted that there is also an expert computer code available to aid in the decision.

Table 2 - Suggested Limits of Different Test Values for Incubated Geomembranes, (see the Reference by Koerner, 1990, for other details and complete references)

(a) For Flexible, Low Crystallinity, Thermoplastic Polymers (Unreinforced and Reinforced), after Little

Property	Resistant	Not Resistant
Permeation Rate (water vapor)	<0.9 $g/m^2/hr$	>0.9 $g/m^2/hr$
Change in weight (%)	<10	>10
Change in volume (%)	<10	>10
Change in tensile strength (%)	<20	>20
Change in elongation at break (%)	<30	>30
Change in 100% or 200% modulus (%)	<30	>30
Change in hardness	10 points	10 points

(b) For Stiff, Semi-Crystalline, Thermoplastic, Polymers

Property	O'Toole		Little		Koerner	
	Resistant	Not Resistant	Resistant	Not Resistant	Resistant	Not Resistant
Permeation Rate (water vapor) (g/m^2-hr)	-	-	<0.9	≥0.9	<0.9	≥0.9
Change in Weight (%)	<0.5	>1.0	<3	≥3	<2	≥2
Change in Volume (%)	<0.2	>0.5	<1	≥1	<1	≥1
Change in Yield strength (%)	<10	>20	<20	≥20	<20	≥20
Change in Yield Elongation (%)	-	-	<20	≥20	<30	≥30
Change in Modulus (%)	-	-	-	-	<30	≥30
Change in Tear Strength (%)	-	-	-	-	<20	≥20
Change in Puncture Strength (%)	-	-	-	-	<30	≥30

(d) Degradation by Swelling - One indication of a geomembrane's durability is the amount of swelling that occurs due to liquid absorption. It should be emphasized that swelling per se does not necessarily mean chain scission nor a failed system. It is, however, a bit disconcerting, and usually results in a change of physical and mechanical properties, at least on a temporary basis.

The test for water absorption, which can be modified for any liquid, is given in ASTM D570. The test is directed at a quantitative determination of the amount of water absorbed, but it is also used as a quality control test on the uniformity of the finished product. The test procedure cautions that the liquid absorption may be significantly different through the edge or through the surface, particularly with laminated products. (This fact alone suggests that in seaming of laminated geomembranes, the upper overlap must be protected against moisture uptake.) Test specimens of 75 by 25 mm are used and immersed in a number of possible ways:

- For 2 hr., 24 hr., or 2 weeks of constant immersion in 23°C water.
- Under cyclic (repeated) immersion.
- For 0.5 hr. or 2 hr. of constant immersion in 50°C water.
- For 0.5 hr. or 2 hr. of constant immersion in boiling water.

The resulting test data are reported as the percentage increase in weight using deionized and distilled water. Some typical values for commonly used geomembranes are as follows (Haxo, Nelson and Miedema 1985):

- PVC – 3 to 4%
- CPE – 1 to 4%
- CSPE – 1 to 5%
- HDPE – negligible
- VLDPE – not evaluated

Swelling due to other liquids is mentioned in the reference cited.

(e) Degradation by Extraction - Some polymers exhibit degradation by the long-term extraction of one or more components of the compound from the polymeric material. These are usually polymers which have been compounded with the use of plasticizers and/or fillers. The as-formulated and compounded mixture of such polymers is very intricate and the bonding mechanism is very complex. When extraction of plasticizers does occur, a sticky surface on the geomembrane results with the remaining structure showing signs of increased modulus and strength, and a lowering of the elongation at failure, i.e. the material becomes progressively brittle (Doyle and Baker 1989). The long term behavior, however, is unknown. It is also possible that anti-degradient components within the polymer may be extracted and leach out to the surface. This might indicate that the remaining polymer is somewhat more sensitive to long-term degradation.

(f) <u>Degradation by Delamination</u> - For geomembranes which are manufactured by calendering or spread coating, delamination is a possibility. It is usually observed when liquid enters into the edge of the geomembrane and is drawn into the interface by capillary tension. This can occur between geomembrane plies, between reinforcing scrim and one of the plies, or between the geomembrane coating of a fabric substrate as in spread coated geomembranes. When it occurs, the individual components are separated and composite action is lost. This type of wicking action has been problematic in the past but current manufacturing methods and proper CQC/CQA in field operations have almost eliminated the situation.

(g) <u>Oxidation Degradation</u> - Whenever a free radical is created, e.g., on a carbon atom in the polyethylene chain, oxygen can create large scale degradation. The oxygen combines with the free radical to form a hydroperoxy radical, which is passed around within the molecular structure. It eventually reacts with another polymer chain creating a new free radical causing chain scission. The reaction generally accelerates once it is triggered as shown in the following equations.

$$R^{\bullet} + O_2 \rightarrow ROO^{\bullet}$$

$$ROO^{\bullet} + RH \rightarrow ROOH + R^{\bullet}$$

where

$R^{\bullet}$ = free radical

$ROO^{\bullet}$ = hydroperoxy free radical

RH = polymer chain

ROOH = oxidized polymer chain

Anti-oxidation additives are added to the compound to scavenge these free radicals in order to halt, or at least to interfere with, the process. These additives, or stabilizers, are specific to each type of resin. This area is very sophisticated and quite advanced with all resin manufacturers being involved in a meaningful and positive way. The specific anti-oxidants are usually proprietary. Removal of oxygen from the geomembrane's surface, of course, eliminates the concern. Thus once placed and covered with waste, or liquid, degradation by oxidation should be greatly retarded. Conversely, exposed geomembranes, or those covered by nonsaturated soil, will be susceptible to the phenomenon.

(h) Biological Degradation - Within the various plant forms of biological life, i.e., bacteria, actinomycetes, fungi and algae, polymer degradation is essentially impossible due to the high molecular weight of the common resins used in geomembranes. In order for such degradation to occur the chain ends must be accessible and this is highly unlikely for molecular weights greater than 1000, let alone 10,000 to 30,000 which is common for geomembrane resins. Biological degradation might be possible for plasticizers or additives compounded with the resin, but information is not authoritative on this subject.

Within the higher forms of biological life, i.e., protozoa, spiders, insects, moles, rats and small mammals, polymers do not contain food and thus are unlikely to be consumed. It is possible, however, that an animal may try to penetrate the geomembrane for access to the opposite side. Here hardness of the predator's teeth enamel versus the geomembrane's hardness is the key comparison. While such events are possible, authoritative information on geomembranes being penetrated in this manner is not known to the authors.

Synergistic Effects

While not degradation mechanisms within, and of, themselves, there are several phenomena which can readily work in conjunction with the previously discussed items. They generally have the effect of accelerating the specific degradation process and thus are called "synergistic effects". At the outset it should be noted, however, that the quantification of these effects is very complicated and the data base is very weak in this regard.

(a) Elevated Temperatures - Whenever the temperature at the surface of, or within, the geomembrane is increased, the mobility of the polymer (and all of its other ingredients) is increased and degradation is usually accelerated. Of the different degradation mechanisms mentioned earlier, this is the case for all of them with the possible exception of biological degradation, which was seen to be of negligible significance. Clearly, elevated temperatures accelerate ultraviolet degradation and this phenomenon has perhaps the largest data base. Thus extreme conservatism is usually taken when testing for ultraviolet degradation. As mentioned earlier, chemical resistance incubation is usually done at an elevated temperature of 50°C for comparison to the 23°C "standard" temperature. Invariably, the higher temperature produces results having greater changes than the lower temperature. There is an upper limit for such temperature testing, however, and that value is based upon polymer modification not representative of realistic behavior. Its value is undoubtedly resin dependent but largely unknown.

For geomembranes placed in the field, high temperatures can generally be avoided by covering them with soil, liquid or another material. Thus the buried environment greatly reduces temperatures in most geomembrane applications. Notable exceptions are surface impoundment and canal liners (above the liquid surface) and for floating covers. For all of these cases simulated testing is absolutely necessary.

(b) Applied Stresses - Invariably, the testing for geomembrane degradation is done on unstressed laboratory samples. Yet, at the very least, the geomembrane will have compressive stresses imposed, and quite possibly tensile stresses as well. What these stresses do to the degradation of the geomembrane as compared to testing in-isolation is quite unknown. The reason for this lack of data is obvious. The cost of experimentation at elevated temperatures is very high in itself and to stress the geomembrane in some simulated form of biaxial stress would generally be cost prohibitive. Yet, some experimentation with stressed geomembranes should be initiated, at least to note the severity of the effect.

(c) Long Exposure - Long exposure includes a multiplicity of effects such as ultraviolet, extraction, oxidation, etc., which can result in synergistic effects beyond each of the previously discussed phenomenon taken individually. For materials as inert as polyethylene, for example, years of exposure at ambient temperature show no indication of any change of properties. For other polymers some changes in surface texture or even in macroscopic properties might occur, but their influence on the geomembrane's behavior is not clear.

The major authoritative data base on long-term aging is available from Matrecon, Inc., (1988), but the steady development of new polymers and compounds makes the situation elusive to say the least. It should be mentioned that a number of landfill owners are beginning to place geomembrane samples (coupons) in retrieveable locations for annual exhuming and evaluation. Such studies will eventually be helpful in assessing actual degradation and aging although the coupons are rarely, if ever, in a stressed condition.

Accelerated Testing Methods

Clearly, the long time frames involved in evaluating individual degradation mechanisms at field related temperatures, compounded by synergistic effects, are not providing answers regarding geomembrane behavior fast enough for the decision making processes of today. Thus accelerated testing, either by high stress, elevated temperatures and/or aggressive liquids, is very compelling. Before reviewing these procedures, however, it must be clearly recognized that one is assuming that the high stress, elevated temperature or aggressive liquids used actually simulates extended lifetimes… an assumption which is not readily substantiated. Thus it *might be* that the test procedures to be described here actually form lower-bound conclusions in predicting degradation, i.e., the results may be minimum values but that is not known with any degree of certainty.

(a) Stress Limit Testing - Focusing almost exclusively on HDPE pipe for natural gas transmission, the Gas Research Institute, the Plastic Pipe Institute and the American Gas Association are all very active in various aspects of plastic pipe research and development. The three above-mentioned organizations, together with Battelle Columbus Laboratories sponsor the Plastic Fuel Gas Symposia which are held on a biennial basis and the resulting Proceedings contain many interesting papers. Stress limit testing in the plastic pipe area has proceeded to a point where there are generally accepted testing methods and standards. ASTM D1598 describes a standard experimental procedure and ASTM D2837 gives guidance on the interpretation of the results of the D1598 test method.

In ASTM D1598, long pieces of un-notched pipe are tightly capped and placed in a constant temperature environment. Room temperature of 23°C is usually used. The pipes are placed under various internal pressures which mobilize different values of hoop stress in the pipe walls, and the pipes are monitored until failure occurs. This is indicated by a sudden loss of pressure. Then the values of hoop stress are plotted versus failure times on a log-log scale, see Figure 2. If the plot is reasonably linear, a straight line is extrapolated to the desired, or design, lifetime which is often 10^5 hours or 11.4 years. The stress at this failure time multiplied by an appropriate factor is called the hydrostatic design basis stress. While of interest for pipelines, the stress state of geomembranes is essentially unknown and is extremely difficult to model. Thus the technique is not of direct value for geomembrane design. It leads, however, to the next method.

(b) Rate Process Method (RPM) for Pipes - Research at the Gas Institute of The Netherlands (Wolters 1987) uses the method of pipe aging that is most prevalent in Europe. It is also an International Standards Organization (ISO) tentative standard currently in committee. Note that other plastic pipe research institutes also are involved in this type of research. The experiments are again performed using long pieces of un-notched pipe which are tightly capped, but now they are placed in various constant temperature environments. So as to accelerate the process, elevated temperature baths up to 80°C are used. Different pressures are put in the pipes at each temperature so that hoop stress occurs in the pipe walls. The pipes are monitored until failure occurs, resulting in sudden loss of pressure. Two distinct types of failures are found; ductile and brittle. The failure times corresponding to each applied pressure are recorded. A response curve is presented by plotting hoop stress against failure time on a log-log scale.

The rate process method (RPM) is then used to predict a failure curve at some temperature other than those tested, i.e., at a lower (field related) temperature than was evaluated in the high temperature tests. This method is based on an absolute reaction rate theory as developed by Tobolsky and Eyring (1943) for viscoelastic phenomenon. Coleman (1956) has applied it to explain the failure of polymeric fibers. The relationship between failure time and stress is expressed in the form as follows:

$$\log t_f = A_0 T^{-1} + A_1 T^{-1} \sigma \tag{1}$$

where

t_f = time to failure

T = temperature

σ = tensile stress on the fiber

A_0 and A_1 = constants

Bragaw (1983) has revised the above model on polymeric fibers and found three additional equations which yield reasonable correlation to the failure data of HDPE pipe. These three equations are as follows:

$$\log t_f = A_0 + A_1 T^{-1} + A_2 T^{-1} P \tag{2}$$

$$\log t_f = A_0 + A_1 T^{-1} + A_2 \log P \tag{3}$$

$$\log t_f = A_0 + A_1 T^{-1} + A_2 T^{-1} \log P \tag{4}$$

where

P = internal pipe pressure proportional to the hoop stress in the pipe

The application of RPM requires a minimum of two experimental failure curves at different elevated temperatures generally above 40°C. The equation which yields the best correlation to these curves is then used in the prediction procedure for a response curve at a field related temperature e.g., 10°C to 25°C. Two separate extrapolations are required, one for the ductile response and one for the brittle response. Three representative points are chosen on the ductile regions of the two experimental curves. One curve will be selected for two points, and the other, the remaining point. This data is substituted into the chosen equation, i.e., Equation 1, 2 or 3, to obtain the prediction equation for the ductile response of the curve at the desired (lower) temperature. The process is now repeated for the predicted brittle response curve at the same desired temperature. The intersection of these two lines defines the transition time.

Figure 3 shows two experimental failure curves which were conducted at temperatures of 80°C and 60°C along with the predicted curve at 20°C. The intersection of the linear portions of the 20°C curve represents the anticipated time for transition in the HPDE pipe from a ductile to a brittle behavior of the material. For pipe design, however, the intersection of the desired service lifetime, say 50 years, with the brittle curve is the focal point. A factor of safety is then placed on this value, e.g., note that it is lowered to 6.5 MPa in Figure 3. This value of stress is used as a limiting value for internal pressure in the pipe.

(c) Rate Process Method (RPM) for Geomembranes - A similar RPM method to that just described for HDPE pipes can be applied to HDPE geomembranes. The major difference is the method of stressing the material. The geomembrane tests are performed using a notched constant load test (NCLT). In this test, dumbbell shaped specimens are taken from the geomembrane sheet. A notch is introduced on one of the surfaces; the notch depth being 20% of the thickness of the sheet. The full description of the

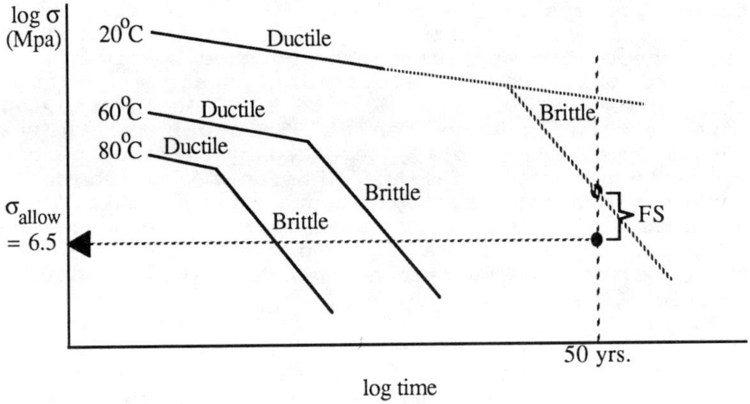

Figure 3 - Burst Test Data for MDPE Pipe. The Intersection of the Ductile Portion of the 20 C Line and 50 Years Has Been Lowered to 6.5 MPa by Multiplication with the Appropriate Factor of Safety. (From Ref. 14)

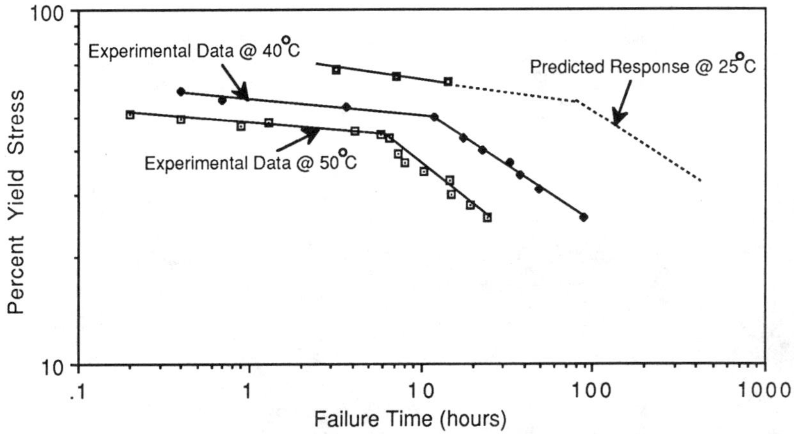

Figure 4 - Notched Constant Load Tests (NCLT) on HDPE Geomembrane Samples Immersed in 10% Igepal/90% Tap Water Solution.

notching process is described by Halse, et al. (1990). Tensile loads varying from 30% to 70% of the yield stress of the sheet, are applied to the notched specimens. The tests are performed in elevated constant temperature environments (usually from 40°C to 80°) and in a surface active wetting agent. Often a 10% Igepal and 90% tap water is used. The data is presented by plotting percent yield stress against failure time on a log-log scale. Figure 4 shows typical experimental curves at 50°C and 40°C which are seen to be very similar to the behavior of HDPE pipe, recall Figure 3. Here distinct ductile and brittle regions can be seen along with a clearly defined transition time.

In order to use these elevated temperature curves to obtain the transition time for a realistic temperature of a geomembrane beneath solid waste or liquid impoundments, e.g., at 25°C, only Equations 3 or 4 can be used due to the data being plotted on a log-log scale. The following is a step-by-step procedure of how the experimental data is used in this regard.

Step. 1 Determine the best fit equation among Equations 3 or 4. We will select Equation 3 which has been found to show the best fit for both the ductile and brittle data.

Step. 2 Obtain the equation for the ductile region of the 25°C curve. To predict the ductile region of the 25°C curve, two data points were chosen from the ductile region of the 50°C curve and one from the 40°C curve. The details of their values are as follows:

Point 1: T_1 = 323 °K; P_1 = 48% = 9.72 MPa; t_1 = 1 h

Point 2: T_2 = 313 °K; P_2 = 52% = 10.53 MPa; t_2 = 5 h

Point 3: T_3 = 313 °K; P_3 = 56% = 11.34 MPa; t_3 = 1 h

Note that the calculations are in degrees Kelvin which is degrees Centigrade plus 273 degrees. By substituting the above three sets of data into Equation (3), and solving the resulting three simultaneous equations, three constants are obtained and are listed below:

$A_0 = -37.18$, $A_1 = 18774$, and $A_2 = -21.2$

By substituting these constants back into

Equation (3) the following equation results for predicting the ductile portion of the curve at any temperature.

$$\log t = -87.18 + 18774/T - 21.2 \log P \quad (5)$$

Step 3. Obtain the equation for the brittle region of the 25°C curve. For the brittle region, two data points were chosen from the brittle region of the 50°C curve and one point from the 40°C curve. The details of their values are as follows:

Point 1: T_1 = 323 °K; P_1 = 37% = 7.50 MPa;
t_1 = 10 h
Point 2: T_2 = 323 °K; P_2 = 28% = 5.67 MPa;
t_2 = 20 h
Point 3: T_3 = 313 °K; P_3 = 37% = 7.50 MPa;
t_3 = 30 h

The resulting three constants obtained by simultaneous solution of the resulting three forms of Equation 3 are listed below:

$A_0 = -11.8$, $A_1 = 4853$, and $A_2 = -2.5$

Therefore, the equation for predicting the brittle portion of the curve at any temperature is as follows:

$$\log t = -11.8 + 4853/T - 2.5 \log P \quad (6)$$

Step 4. Use Equations (5) and (6) to obtain the predicted response curves at the desired temperature. The ductile behavior from Equation (5) and the brittle behavior from Equation (6) at the specific temperature of 25°C was used to plot the desired curves at 25°C in Figure 4.

Step 5. Intersect the two prediction curves for the ductile-to-brittle transition time. As seen in Figure 4 this intersection produces a transition time for the 25°C temperature response of 60 hours. Note that the ductile-to-brittle transition time is increased over ten fold (from 6 to 70 hours) by decreasing 25 degrees in temperature. It should also be noted that this curve represents the behavior of the geomembrane in 10% Igepal solution. The performance of the same geomembrane will probably be very different in other environments, such as in leachate or water.

(d) Elevated Temperature and Arrhenius Modeling - Using an experimental chamber as shown in Figure 5, Mitchell and Spanner (1985) have superimposed compressive stress, chemical exposure, elevated temperature and long testing time into one experimental device. For their particular tests, three duplicate chambers were operated at 18°C, 48°C and 78°C respectively. At the end of the arbitrarily designated test period (in the example to be described it was 18 weeks), the geomembrane samples were removed. Mechanical tests and chemical analyses were performed on these incubated samples to monitor if any changes in the various properties of the geomembranes occurred.

The mechanical tests included the following:

- Tensile Strength and Elongation
- Yield Strength and Elongation
- Stress Cracking Behavior

The chemical analyses included the following:

- Differential scanning calorimetry (DSC); for measuring crystallinity and oxidation induction time (OIT)
- Infra-red spectrometry (IR); for measuring concentration of carbonyl groups
- Gel permeation chromatography (GPC); for measuring the molecular weight and molecular weight distribution

If there were changes in any of the above properties; for example, in the concentration of the carbonyl group, the reaction rate (K) was obtained for each experimental test temperature (T). These values were now used with the Arrhenius equation which is as follows (American National Standard 1986):

$$K = Ae^{-E/RT} \tag{4}$$

where

K = reaction rate for the process considered
A = a constant for the process considered
E = reaction activation energy for the process considered
T = temperature (°K =°C + 273)
R = gas constant (8.314 J/mol-°K)

By plotting "ln K" against "1/T", a straight line should be obtained, see Figure 6. The slope of this line is "-E/R" for the particular property change being

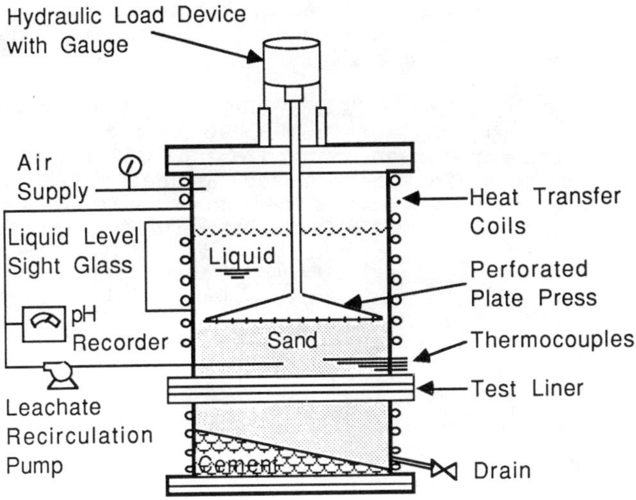

Figure 5 - Schematic of Accelerated Aging Column (after Mitchell and Spanner)

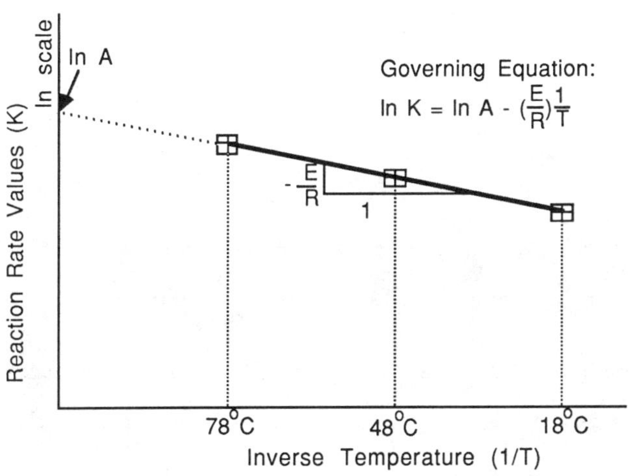

Figure 6 - Graphical Method of Plotting Reaction Rate Values for Change in a Specific Geomembrane Property.

monitored. The constant "A" can also be identified but it drops out of the equation when comparing the responses at two different temperatures.

For example, Turi (1981) found that the activation energy (E) of polyethylene is 109,000 J/mol. This is the energy requirement for an atom to change its position. If we use this value in Mitchell and Spanner's tests, then the degradation that occurs between the two tests at temperatures of 18°C and 78°C is determined as follows:

$$\frac{r_{78\,+\,273}}{r_{18\,+\,273}} = e^{-\frac{(109{,}000)}{8.314}\left[\frac{1}{351} - \frac{1}{291}\right]}$$

$$= 2254$$

Hence, for their particular process occurring in 18 weeks at 78°C it would take 2264 times 18 weeks, or 784 years, to occur at 18°C.

(e) Hoechst Multiparameter Approach - The Hoechst research laboratory in West Germany has been active in long term testing of HDPE pipe since the 1950's. Recently they have been applying their expertise and experience to the long term behavior of HDPE geomembranes (Kork, et al. 1987). Note, however, that there are major differences between pipe and sheet. The stress state in pressurized pipe is well known, whereas the stress state in the geomembranes in the field is not nearly as well known. Furthermore, the pipe is under a constant stress state, whereas stress relaxation can occur in geomembranes. If these two differences in the state of stress between pipe and sheet can be resolved, then the tremendous wealth of data obtained in over 30 years of pipeline testing can be carried over to the geomembrane area.

The Hoechst group has considered geomembranes for the case of local subsidence under the liner and the mobilization of out-of-plane deformation of the sheet into an multi-axial stress state. Thus their model for sheet stresses is biaxial stress. (In the usual pressurized pipe, the radial stress is twice the value of the longitudinal stress. Hence the isotropic biaxial stress state in a normal pipe testing experiment was achieved by putting the pipes under an additional longitudinal stress. It was found that the lifetimes in these tests were the same as in normal pressure burst testing. Hence if biaxial stress relaxation could be accounted for, a viable long term testing technique could very well be developed for sheet from modified pipe testing. This of course assumes that the different manufacturing and processing of pipe and sheet produce essentially the same material properties. (This may not be completely the case.) The Hoechst long term testing for geomembrane "sheet" thus consists of the following procedure:

(1) Modified burst testing of pipe (of the same material as the sheet), with additional longitudinal stress to produce an isotropic, biaxial stress state. Their tests make note that the site-specific liquid should be used.

(2) Assume a given subsidence strain versus time profile.

(3) Measurement of stress relaxation curves in sheets which have been stressed biaxially, at strain values encountered in field.

(4) Use steps (2) and (3) to predict the stress as a function of time.

(5) See how these maximum stresses compare with the stress-lifetime curves determined in the normal constant stress-lifetime pipe measurements of Step (1). The constant stress-lifetime curves are modified (as in the normal pipe testing) to accommodate the effects of various chemicals and even for seams.

(6) If linear accumulation of degradation is assumed, then the variable stress curves can be used to predict failure from the constant stress-lifetime curves.

The steps (1)-(5) have been performed by Kork, et al. (1987). The approach should certainly be considered seriously. One of the main impediments to its viability would seem to be that if one produces pipe, the final product may have different material properties than sheet. For example, the residual stresses could be quite different. Other work of a related nature can be found in Hessell and John (1987) and Gaube, et al. (1976).

Summary and Conclusions

This rather lengthy treatise on durability and aging has attempted to give insight into long-term performance of geomembranes by itemizing those mechanisms which can degrade the polymer resin and/or compound. Table 3 gives a summary of the individual degradation mechanisms that were discussed. All of them, taken individually, are possible to evaluate and/or quantify. In addition, some suggestions as to preventative measures are offered.

Table 3 - Degradation Phenomena In Geomembranes (from a Geosynthetic Engineering Perspective)

Degradation Classification	Degradation Mechanism	Initial Change in Material		Subsequent Change in Material		Preventative Measure
		Laboratory(1)	Field(2)	Laboratory(3)	Field(4)	
Ultraviolet	• chain scission • bond breaking	• mol. wt. • stress crack resist.	• color • crazing	• elongation • modulus • strength	• color • cracking	• screening agent • anti-degradient • cover geomembrane
Radiation	• chain scission	• mol. wt. • stress crack resist.	• color • crazing	• elongation • modulus • strength	• color • cracking	• cover for β and α-rays • shield for neutrons • reduce dosage for γ rays
Chemical	• reaction with structure • reaction with additives	• carbonyl index • IR • mol. wt.	• texture • color • crazing • TGA	• elongation • modulus • strength	• texture • cracking • reactions	• proper resin • proper additives
Swelling	• liquid absorption	• TGA	• thickness • color • texture	• thickness • modulus • strength	• thickness • softness	• proper resin • proper manufacturing
Extraction	• additive expulsion	• TGA • IR	• texture • color • thickness	• elongation • modulus • thickness	• texture • color	• proper compounds • proper manufacturing

Table 3 - Continued

Degradation Classification	Degradation Mechanism	Initial Change in Material: Laboratory[1]	Initial Change in Material: Field[2]	Subsequent Change in Material: Laboratory[3]	Subsequent Change in Material: Field[4]	Preventative Measure
Delamination	• adhesion loss	• thickness • edge effects	• thickness • edge effects	• thickness • ply adhesion	• layer separation • thickness	• proper manufacturing • protect geomembrane edges
Oxidation	• reaction with structure	• IR • carbonyl index • OIT	• color • crazing	• elongation • modulus • strength	• color • cracking	• anti-oxidant • cover with soil • cover with liquid
Biological	• reactions with additives	• mol. wt. • carbonyl index • IR	• texture • surface film	• elongation • modulus • strength	• texture • surface layer • cracking	• avoid sensitive additives • biocide

Notes:

(1) Initial laboratory changes are generally sensed by chemical fingerprinting methods: mol. wt. = molecular weight, IR = infrared spectroscopy, TGA = thermal geometric analysis, OIT = oxidation inducation time

(2) Initial field changes are generally sensed on a qualitative basis.

(3) Subsequent laboratory changes can be sensed by numerous physical and mechanical tests. Listed in the table are those considered to be the most sensitive parameters.

(4) Subsequent field changes are a continuation of the initial changes until physical and mechanical properties being to visually change.

Also described in the paper were various synergistic effects which greatly complicate the above mechanisms when acting concurrently. Items such as elevated temperature, biaxial or triaxial stress and long exposure time are very difficult to model accurately. Nevertheless, much laboratory modeling had been done, although most has been by the plastic pipe industry. This is almost exclusively on polyethylene pipe and the technology transfer to polyethylene geomembranes is certainly very valuable. Many of the techniques are being evaluated for long-term geomembrane performance; for example, ductile-to-brittle transition time along with Arrhenius modeling. Other types of geomembranes appear to be in need of long-term simulation testing as was discussed in various facets of the paper.

In conclusion, it is felt that case histories (both positive and negative) give the best insight into the field behavior of geomembrane lined facilities at this point in time. Test strips which are exhumed annually and tested, versus the original material are extremely valuable. They can be placed along the edge of the facility or within sump areas for easy removal. Field failures are also very important to analyze for aiding and prompting in the modification of existing polymer formulations and perhaps changing the basic resins themselves. Lastly, long term laboratory tests under simulated conditions should be undertaken. Simulated stress tests under elevated temperature testing and Arrhenius modeling is very intriguing in this regard as is some type of modification of the Hoechst procedures. Answers to the important question of "how long will they last" may never be known unless the effort begins as soon as possible.

References

-----, U. S. Federal Register, Regulations on Liner Systems, July 26, 1982.

-----, U. S. EPA Method 9090, Compatibility Test for Wastes and Membrane Liners, in Test Methods for Evaluating Solid Waste Physical/Chemical Methods, SW-846, 2nd Edition, 1984.

-----, U.S. EPA Computer Code on Flexible Membrane Liner Advisory Expert System (FLEX), Cincinnati, Ohio, 1987.

-----, "Standard for Polymeric Materials - Long Term Property Evaluations," American National Standard, ANSI/UL 746B - 1986.

Apse, J. I., "Polyethylene Resins for Geomembrane Applications," Durability and Aging of Geosynthetics, R. M. Koerner, Ed., Elsevier Appl. Sci. Publ. Ltd., 1989, pp. 159-176.

Bragaw, G. G., "Service Rating of Polyethylene Piping Systems by the Rate Process Method," 8th Plastic Fuel Gas Pipe Symp., Nov. 1983.

Charlesby, A., Atomic Radiation and Polymers, Oxford University Press, 1960.

Chuck, R. T., "Largest Butyl Rubber Lined Reservoir," Civil Engineering, ASCE, May 1970, pp. 44-47.

Coleman, B. D., "Application of the Theory of Absolute Reaction Rates to the Creep Failure of Polymeric Filaments," Jour. Polymer Science, Vol. 20, 1956, pp. 447-455.

Doyle, R. A. and Baker, K. C., "Weathering Tests of Geomembranes," Durability and Aging of Geosynthetics, R. M. Koerner, Ed., Elsevier Appl. Sci. Publ. Ltd., 1989, pp. 152-158.

Gamski, K., "Geomembranes: Classification, Uses and Performance," Jour. Geotex. and Geomemb., Elsevier Appl. Sci. Publ. Ltd., Vol. 1, 1984, pp. 85-117.

Gaube, E., Diedrick, G. and Muller, W., "Pipes of Thermoplastics; Experience of 20 Years of Pipe Testing, Kunstoffe, Vol. 66, 1976, pp. 2-8.

Geier, F. H. and Morrison, W. R., "Buried Asphalt Membrane Canal Lining," Research Report No. 12, A Water Resources Technical Publication, Bureau of Reclamation, Denver, Colorado, 1968.

Grassie, N. and Scott, G., Polymer Degradation and Stabilization, Cambridge University Press, 1985.

Halse,Y. H., Lord, A. E. Jr. and Koerner, R. M., "Ductile-to-Brittle Transition Time in Polyethylene Geomembrane Sheet," Symposium on Geosynthetic Testing for Waste Containment Applications, ASTM Spec. Tech. Publ., to appear in 1990.

Haxo, H. E., Nelson, N. A. and Miedema, T. A., "Solubility Parameters for Predicting Membrane Waste Liquid Compatibility," Proc. EPA Conf. on Hazardous Waste, Cincinnati, OH, Apr. 1985, pp. 198-215.

Hessel, J. and John, P., "Long Term Strength of Welded Joints in Polyethylene Sealing Sheets," Werkstofftechnik, Vol. 18, 1987, pp. 228-231.

Kays, W. B., Construction of Linings for Reservoirs, Tanks and Pollution Control Facilities, 2nd Ed., J. Wiley and Sons, Inc., NY, 1988.

Koerner, R. M., Designing with Geosynthetics, 2nd Ed., Prentice Hall Publ. Co., Englewood Cliffs, NJ, 1990.

Kork, R., et. al., "Long Term Creep Resistance of Sheets of Polyethylene Geomembrane, Report TR-88-0054 from Hoechst A.G., Frankfurt, W. Germany, 1987.

Matrecon, Inc., Lining of Waste Impoundment and Disposal Facilities, EPA/600/2-88.052, Sept., 1988.

Mitchell, D. H and Spanner, G. E., "Field Performance Assessment of Synthetic Liners for Uranium Tailings Ponds," Status Report, Battelle PNL, U.S. NRC, NUREG/CR-4023, PNL-5005, Jan., 1985.

Phillips, D. C., "Effects of Radiation on Polymers," Materials Science and Technology, Vol. 4, 1988, pp. 85-91.

Staff, C. E., "The Foundation and Growth of the Geomembrane Industry in the United States," Intl. Conf. Proc. on Geomembranes, Dever, Colorado, IFAI, 1984, pp. 5-8.

Tobolsky, A. and Eyring, H., "Mechanical Properties of Polymeric Materials," Jour. Chem. Phys., Vol. 11, 1943, pp. 125-134.

Turi, A., Thermal Characterization of Polymeric Materials, Academic Press, 1981.

Van Zaten, R. V., Geotextiles and Geomembranes in Civil Engineering, A. A. Balkema Press, Rotterdam and Boston, 1986.

Whyatt, G. A. and Farnsworth, R. K., "The Effect of Radiation on the Properties of HDPE and PP Liners," Symposium on Geosynthetic Testing for Waste Containment Applications, ASTM Spec. Tech. Publ., to appear in 1990.

Wolters, M., "Prediction of Long-Term Strength of Plastic Pipe," Proc. 10th Plastic Fuel Gas Symp., 1987, New Orleans, Amer. Gas Assoc. Publ.

Acknowledgements

The authors would like to express sincere appreciation to all of the sponsoring member organizations of the Geosynthetic Research Institute. A listing of the firms and their contact members is available from the authors. The Institute is focused on long term generic research in geosynthetics of the type described in this paper.

Construction and Performance of a Long-Term Earthen Liner Experiment

Keros Cartwright[1]
Ivan G. Krapac[2]

Abstract

In land burial schemes, compacted soil barriers with low hydraulic conductivity are commonly used in cover and liner systems to control the movement of liquids and prevent groundwater contamination. An experimental liner measuring 8 x 15 x 0.9 m was constructed with design criteria and equipment to simulate construction of soil liners built at waste disposal facilities. The surface of the liner was flooded with a 29.5 cm deep pond on April 12, 1988. Infiltration of water into the liner has been monitored for two years using 4 large-ring (1.5 m OD) and 32 small-ring (0.28 m OD) infiltrometers, and a water-balance that accounts for total infiltration and evaporation. Average long-term infiltration fluxes based on two years of monitoring are 5.8×10^{-9} cm/s, 6.0×10^{-8} cm/s and 5.6×10^{-8} for the large-ring, small-ring, and water-balance data, respectively. The saturated hydraulic conductivity of the liner based on small-ring data, estimated using Darcy's Law and the Green-Ampt Approximation, is 3×10^{-8} and 4×10^{-8} cm/s, respectively. All sets of data indicate that the liner's performance exceed that which is required by the U.S. EPA.

---

[1]Keros Cartwright, Principal Geologist, Illinois State Geological Survey, 615 East Peabody Drive, Champaign, IL 61820.

[2]Ivan Krapac, Staff Geochemist, Illinois State Geological Survey, 615 East Peabody Drive, Champaign, IL 61820.

Tensiometer and water-balance data indicate that the depth of the wetting front is approximately 44 cm below the surface of the liner after two years of monitoring. Based on the depth of the wetting front and apparent rate of movement, breakthrough should occur approximately six years (1994) after ponding of the liner.

Introduction

Compacted soil liners are widely used at landfill and waste lagoon facilities as a barrier between the waste materials and the hydrogeologic environment. When the landfill or lagoon contains wastes classified as hazardous, the facility is required to have a double liner system and leachate collection and leak detection systems, all underlain by a compacted soil liner. Municipal refuse disposal is generally subject to less stringent requirements than hazardous waste, and naturally occurring or recompacted soil materials may be used as the liner. The Resource Conservation and Recovery Act (RCRA) requires that the earthen portion of the liner must be a minimum of 0.9 meters thick and have a saturated permeability of no more than 1×10^{-7} cm/s (U.S. Federal Register, 1987).

Although soil liners are commonly used, research results from field-scale soil liners to evaluate their performance are limited (U.S. EPA, 1988). Existing detailed field-scale experiments of soil liners have been conducted by Rogowski (1972, 1989) and Elsbury et al. (1988). Neither study produced a liner hydraulic conductivity that met the maximum U.S. EPA requirement (not necessarily a project goal); hydraulic conductivities obtained were 5×10^{-7} (Rogowski, 1989) and 1×10^{-4} cm/s (Elsbury et al., 1988). Daniel and Brown (1988) compiled a table of 14 case studies of earthen liners constructed throughout the United States and Canada. Of these 14 studies, only two met the U.S. EPA hydraulic conductivity requirement of 1×10^{-7} cm/s or less. The two successful liners were the Keele Valley Landfill in Ontario (Daniel and Trautwein, 1986), and a test liner in California (Chen and Yamamoto, 1987). The construction of both liners utilized rigorous construction quality assurance plans. Daniel and Brown (1988) suggested five common causes for failure to meet the hydraulic conductivity criteria: 1) the liners were too thin; 2) improper construction, that is, poor material selection and/or inadequate design; 3) inadequate construction quality assurance; 4) desiccation and freezing; and 5) attack by chemical constitutes in the waste.

In response to the uncertainty of liner performance, the Illinois Geological Survey initiated a study in September, 1985, of the performance of a field-scale soil liner using *in situ* measurements of construction (e.g., soil density and moisture content) and hydrogeologic parameters (e.g., infiltration rates, moisture movement). Specific overall objectives of this study are:

1) Determine whether the U.S. EPA requirement of a saturated hydraulic conductivity of 1 x 10^{-7} cm/s or less can be achieved using *in situ* measurements;

2) Quantify the areal variability of liner hydraulic properties;

3) Determine the transit times of water and tracers through a compacted soil liner;

4) Test the accuracy of methods for predicting transit time of water and tracers through the liner.

To fulfill these objectives, the study was divided into three phases. Phase 1, which began in September 1985, addressed the evaluation of soil properties suitable for constructing a liner, and the characterization and selection of a soil for use in this project. Phase 2, which began in October 1986, provided for the construction of a prototype liner to test construction practices and determine if the hydraulic conductivity requirement could be obtained using the soil selected. Phase 3 began in October 1987, with the construction and instrumentation of a field-scale soil liner. This liner is extensively instrumented and is currently being monitored. Excellent literature reviews pertaining to construction and evaluation of liners are available in U.S. EPA (1979, 1983, 1988) and Rogowski (1989).

Studies of Fine-Grained Materials

Fine-grained materials composed of a significant percentage of clay-sized particles tend to exhibit the two most desirable properties a liner for a waste disposal facility should possess; a low hydraulic conductivity and a high adsorption capacity. This project only addressed the first of these properties--the low hydraulic conductivity. To predict the behavior of a low-permeability earthen liner, it is particularly important to understand the hydrologic and engineering properties of clay-rich sediments. Detailed descriptions of the properties of clay minerals and clay-sized particles have been published by Grim (1962, and 1968), Mitchell (1976), van Olphen (1963), and Young and Warkentin (1975). More recently, Ely et al. (1983) completed a study of the performance of clay caps and liners used for disposal facilities.

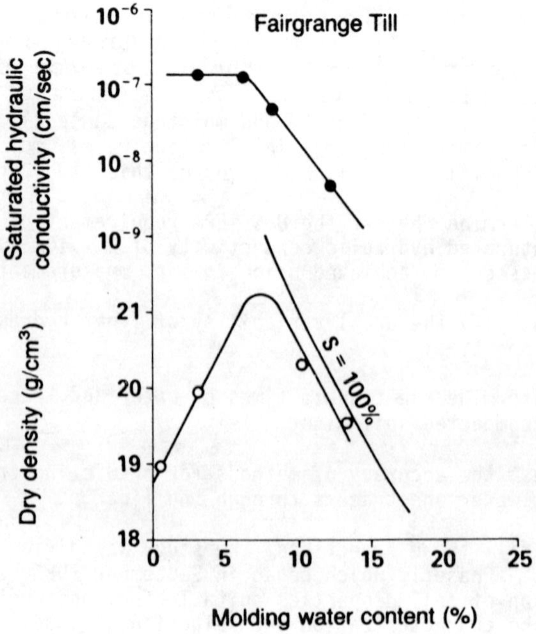

Figure 1. Results of compaction and saturated hydraulic conductivity tests for the Fairgrange Till showing decreases in hydraulic conductivity for samples compacted at water contents greater than optimum. The Fairgrange Till is similar to the Batestown Till used in this experiment (from Cartwright et al., 1988).

The soil property of greatest concern generally is the saturated hydraulic conductivity. This property directly affects the rate of water and contaminant movement through an earthen liner. It has been shown that the hydraulic conductivity of compacted earth materials is significantly affected by the method used to compact the soil and by the water content of the soil (see fig. 1) at the time of compaction (Cartwright et al., 1988; Johnson et al., 1983; Mitchell et al., 1965; Seed and Chan, 1959; Lambe, 1958).

While time and monetary constraints frequently dictate that hydraulic conductivity tests be conducted in the laboratory, there has been considerable discussion regarding the relevance of laboratory tests of hydraulic conductivity to field conditions. The sources of potential error in laboratory tests of hydraulic conductivity have been discussed by Olson and Daniel (1979). These include: 1) compaction and water content higher in the laboratory samples than in the field; 2) air in laboratory samples; 3) use of excessive hydraulic gradient in the laboratory causing particle migration, and; 4) too small a sample size. Other workers have also addressed the issue of confidence in laboratory-derived results of hydraulic conductivity of soils related to size and number of samples (Mason et al., 1957; Daniel, 1981; Anderson and Bouma, 1973). The importance of hydraulic gradient in laboratory tests of hydraulic conductivity has been investigated by Acar and Field (1983) and Zimmie et al. (1981). Also, the ability to achieve saturation of samples has long been of concern (Smith and Browning, 1942; Johnson, 1954).

Studies by Dunn and Mitchell (1984) suggested that present laboratory methods may yield variations of hydraulic conductivity up to several orders of magnitude for a single soil type, and propose a test procedure that would yield consistent results. Daniel et al. (1984) describe a method using a flexible-wall permeameter to eliminate problems of sidewall leakage, while the relative merits of both flexible- and fixed-well permeameters are discussed by Daniel, Anderson and Boynton (1984).

In addition, increasing evidence shows that laboratory tests tend to greatly underestimate hydraulic conductivity when compared to field data for the same soil (Olson and Daniel, 1981; Daniel, 1984; Herzog and Morse, 1986; and U.S. EPA, 1988). The precise reason for this phenomenon is not known; however, it most likely is related to inhomogeneities in the soil. Because of this, it is now generally recognized that the results of laboratory tests should be used to compare the relative hydraulic conductivities of materials, and then correlated with field data for the same material.

Prototype Liner

Selection of Construction Materials

The soil properties assigned maximum or minimum criteria were during the selection process: hydraulic conductivity, Atterberg limits, particle size distribution, natural moisture content, and dry bulk density. The Atterberg limits (plasticity index) criterion could not be met by any of the borrow sources, and had to be relaxed slightly; however, additional testing of the actual materials used in construction did achieve a plasticity index greater than the desired 10 percent. All of the remaining criteria were achieved by the soils. The other soil properties tested (dispersitivity, clay mineralogy, specific gravity, cation exchange capacity) did not provide significantly different results to serve as selection criteria. The final selection decision was based on economic factors because of the similarity in soil properties for the potential liner materials. A paper is in preparation (Su et al.) describing the material testing and selection.

Construction and Results

The prototype liner with the dimensions of 3m x 9m x 0.9m was constructed in a series of six lifts using a pad-foot (10 cm long) compactor (Hyster C862A) capable of delivering 220 kilonewtons of force in a vibratory mode. Based on the soil properties of the liner (moisture content and density), the liner appeared to be relatively homogeneous. However, the moisture content was more variable than density. Densities of 90% of maximum as determined by Standard Proctor were achieved at moisture contents approximately 1 to 2% above optimum.

Hydrologic monitoring of the prototype liner consisted of double-ring infiltrometers (1.5m OP), suction-vacuum lysimeters, and tensiometers. Infiltration rates were measured for a 50-day period. A "steady state" infiltration flux of 1.5×10^{-7} cm/s was achieved two to three weeks after initiation of the infiltration experiment. The calculated saturated hydraulic conductivity was no greater than 3.6×10^{-8} cm/sec (Albrecht and Cartwright, 1989). The wetting front depth, 7.3 cm and 9.6 cm, was calculated for the two large-ring infiltrometers after 46 days of infiltration (Albrecht and Cartwright, 1989). Dyes (Fluorescein and Rhodamine WT) added to the rings of one infiltrometers as tracers were observed when the prototype liner was excaved; the dyes showed a sharp front at 4 cm (Albright et al., 1989). Dye patterns in the liner indicated lateral flow had occurred between lifts and vertical flow had occurred along clod-clod interfaces. The total depth of penetration by the dye along preferential pathways was 15 cm vertically, and dye discharged from the side of the experimental liner approximately one meter from the dye source (Albrecht et al., 1989).

Field Scale Liner

Design

The field-scale liner design incorporate the experience from the prototype liner. The liner was constructed using standard engineering methods and incorporates instrumentation to evaluate the liner at various scales of measurement. The liner design provides for a 10 m x 17 m x 1 m area of compacted soil of which a 8 m x 15 m x 0.9 m test area is instrumented and ponded. The cross-sectional profile and plan view of the liner are shown in Figure 2 and 3. Details of the construction are given in Krapac et al. (1989).

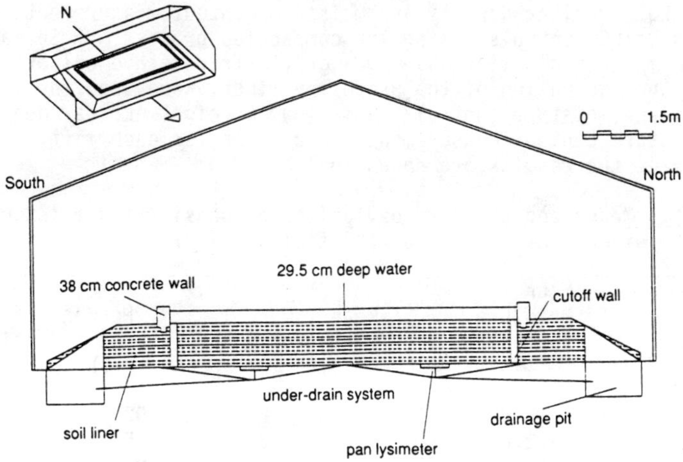

Figure 2. Cross sectional view of the compacted soil liner, shelter, and accessory components (from Krapac et al., 1989).

The underdrain system is divided into four quadrants (Fig. 3), and will serves to collect water that migrates downward through the liner. The volume of water collected in the underdrain system will allow an additional estimate the saturated hydraulic conductivity at the scale of the overall liner. Four pan lysimeters are installed in the underdrain directly beneath the large-ring infiltrometers. There is direct contact between the bottom of the liner and lysimeter providing a minimal "dead space." Water and tracers migrating through the liner can directly move to the lysimeter with little retention, increasing the resolution of breakthrough to the underdrain.

Liner Construction

The liner was constructed by compacting seven lifts. The first lift had an uncompacted (loose) thickness of 31 cm to protect the underlying-underdrain system and pan lysimeters. Each remaining lift had an uncompacted thickness of 23 cm. The material was spread using either the scoop of the compactor or a Harvard Rotavator HR-20 tiller. Each lift was spread, retilled, and then compacted to a thickness of 15 cm. The total compacted thickness of the 7 lifts was 1.1 m. The top 0.2 m was then removed to provide a uniform surface for the experiment.

The compaction specifications required that each lift be compacted with a Caterpillar 815B compactor, a static load pad-foot compactor with 20 cm feet and a rated operating weight of

20,037 kg. Nuclear density and moisture content measurements were used as QA/QC controls during the compaction process. A Seaman nuclear meter and a microwave oven gravimetric method calibrated during the excavation of the prototype liner, were used for density and moisture control. A minimum of eight nuclear density and moisture content measurements were taken for each lift. A summary of the results are shown in Table 1.

Table 1. Means and standard deviations of density and moisture contents of compacted lifts of the field-scale liner.

Lift No.	n	Mean Moisture Content (%)	S	Mean Density gm/cm^3	S	Percent of Proctor Density (m=9.9%)	(m=11.3%)
1	11	11.9	0.7	1.83	5.0	92	95
2	8	12.0	0.7	1.88	3.4	95	97
3	8	10.8	0.9	1.78	4.4	90	92
4	9	10.9	0.8	1.80	3.4	90	93
5	10	11.5	1.0	1.88	3.8	95	97
6	9	11.2	1.0	1.84	5.8	93	95
Overall	55	11.4	1.0	1.84	4.8	93	95

n = Sample size
S = Standard Deviation
m = Moisture content at compaction

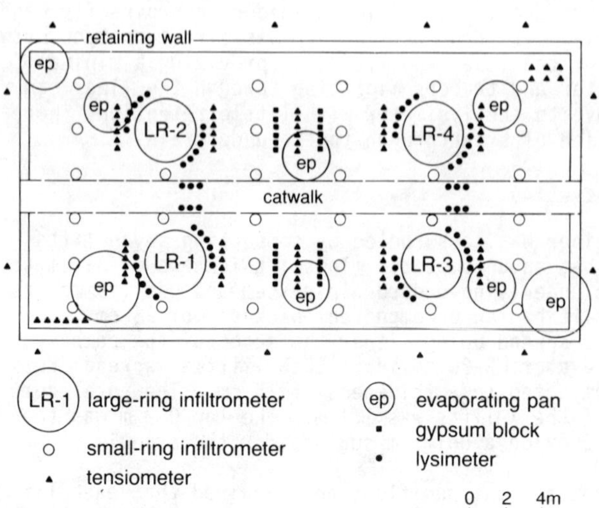

Figure 3. Instrumentation design of the soil liner (from Krapac et al., 1990).

Instrumentation

Infiltration rates are measured using large- (1.5 m) and small-ring (0.28 m) infiltrometers and a water balance for the pond. Other instrumentation includes tensiometers, gypsum blocks, suction-vacuum lysimeters, and ambient environmental monitors (Fig. 3). Bromide, m-TFMBA, o-TFMBA and PFBA were added to the large-ring infiltrometers at the beginning of the experiment to monitor solute movement through the liner. Tritiated water was added to the pond July 20, 1989, 436 days after initial ponding.

The gypsum block (24 total) proved ineffective in monitoring moisture changes and have been abandoned. The 60 suction-vacuum lysimeters installed to monitor tracer movement have as yet, not provided useful data.

Sixty tensiometers are installed in the liner to monitor the movement of moisture and changes in hydraulic head. The tensiometers are placed in ten nests with six tensiometers per nest to measure soil tension at depths of 10, 18, 33, 51, 69, and 89 cm (Fig. 3). Two years after the liner was flooded, 55 of these tensiometers are operational. A complete description of the tensiometer, installation techniques, and operation can be found in Krapac et al. (1989).

Results

Infiltration Measurements

A water-balance is used to determine the overall infiltration flux into the liner for the two-year period between April, 1988 and April, 1990. The volume of water lost to evaporation is subtracted from the volume of water required to maintain a constant level in the pond; the resultant volume (5478.8 L) is assumed to have infiltrated into the liner. On the basis of these data, the infiltration flux for the entire liner is estimated to be 8.4×10^{-8} cm/s. This estimate is an average of both initial and steady-state infiltration flux of the liner; using a least-square regression of the cumulative-infiltration data for the last 730 days (two years) the flux is 6.4×10^{-8} cm/s ($r^2 = 0.97$).

An estimate of the depth of the wetting front two years after ponding can be calculated from the bulk density (1.84 g/cm^3) and initial moisture content (0.21 cm^3/cm^3) of the liner. Using an area of infiltration of 1.3×10^6 cm^2, 5478 liter of water infiltrated in two years, a total porosity of 33% with 21% of the pores initially filled with water, and assuming a uniform wetting front, the wetting front is calculated to be 44 cm below the liner surface, or approximately half way through the liner. Thus, the breakthrough of water is at least four years after ponding, and present data suggest about six years.

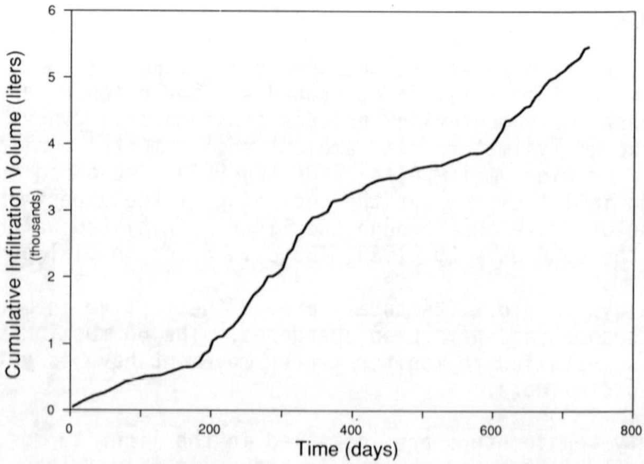

Figure 4. Cummulative infiltration from April 1988 to April 1990 for the entire liner (after Panno et al., 1990).

A three-fold periodic change in infiltration rate for the entire liner is observed in the cumulative-infiltration from the pond (Fig. 4). There is an increase in infiltration rate followed by a decrease and then an increase which occurred in mid-October, mid-March, and mid-October, respectively. These changes correspond to the seasons and approximately the time when the liner shelter's heater cycled on and off. This relationship suggests that the infiltration rate fluctuations may be temperature related. Jaynes (1990) observed a cyclical variation in the infiltration rate of ponded soil as a result of diurnal changes in soil temperature. It is also possible that the eight evaporation pans underestimate evaporation when the shelter's heater is in use. These relationships are currently under investigation.

The infiltration-flux data (representing steady-state infiltrability) from the small ring (SR) and large ring (LR) infiltrometers form two statistically distinct populations (Fig. 5a,b). The SR-infiltrometer fluxes (the distribution on the left side of the diagram) are log-normally distributed on the basis of the Kolmogorov-Smirnov test at the 95% confidence level (Panno et al., 1990). This distribution is consistent with work by Rogowski (1972), Nielson et al. (1973), and Parkin et al. (1988) who reported the lognormal spatial distribution of soil hydraulic properties. The mean infiltration flux and variance based on the

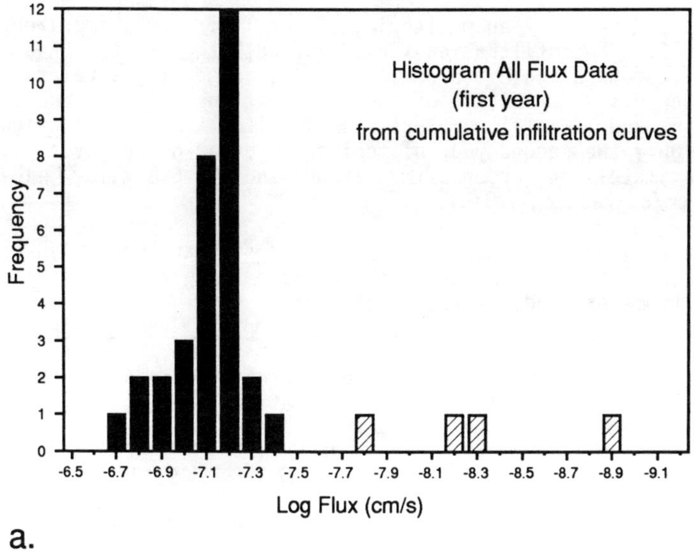

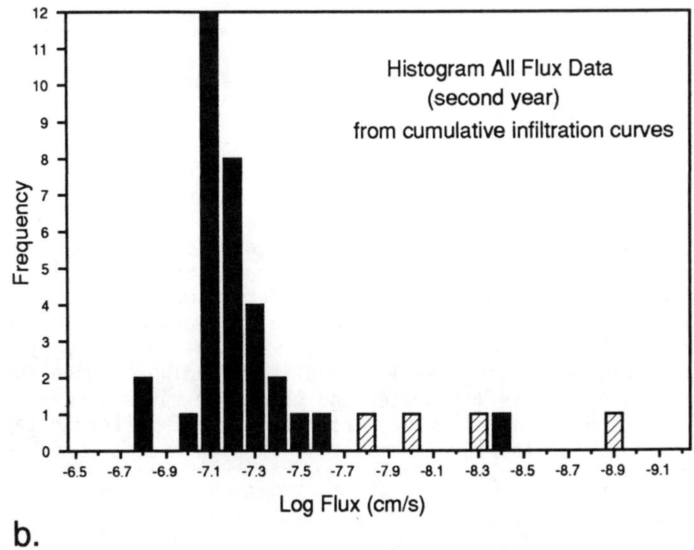

Figure 5. Historgrams of log-transformed fluxes measured by the SR- and LR-infiltrometers after: a) one year of infiltration; b) two years of infiltration (after Panno et al., 1990).

first and second year monitoring of the liner are summarized in Table 2. The infiltration fluxes for both the LR- and SR-infiltrometers exhibit a small variance suggesting a relatively homogeneous distribution of the infiltration flux over the liner. The LR- and SR-infiltrometers mean infiltration fluxes, calculated from only the second year of monitoring data, differ by approximately one order of magnitude, and are 5.8×10^{-9} and 6.0×10^{-8} cm/s, respectively.

Table 2. Geometric mean, Log mean, and standard deviation of infiltration fluxes measured by the large-ring and small-ring infiltrometers and the water balance.

Instrument	Infiltration Flux (cm/s)	Log Mean of Infiltration	Standard Deviation*	Sample Size
LR (1st Year)	5.0×10^{-9}	-8.301	0.458	4
LR (2nd Year)	5.8×10^{-9}	-8.237	0.474	4
SR (1st Year)	8.0×10^{-8}	-7.100	0.152	31
SR (2nd Year)	6.0×10^{-8}	-7.218	0.268	32
Water balance	6.4×10^{-8}			1

*The log mean plus or minus 2 times the log standard deviations corresponds to a range of infiltration flux values from 6.1×10^{-10} to 1.6×10^{-7} cm/s.

The reason for the difference between the SR- and LR-infiltrometer fluxes is unclear; however, a possible explanation is leakage of water around or through the grout seal of the LR-infiltrometers. Evidence for this leakage includes the observations of gas bubbles emanating from the grout-soil interface outside and inside the ring during maintenance operations. The gas bubbles indicate a connection between the inside of the LRs and the liner pond. Excavation of the prototype liner (Albrecht et al., 1989) showed that bonding between the fiberglass of the infiltrometer and the liner material was imperfect and a possible source of the leakage. Although the magnitude of the leakage is small, it is significant compared to the very low infiltration flux.

Hydraulic Conductivity and Transit Time

The saturated hydraulic conductivity can be estimated for the liner using measurements of steady-state infiltrability and hydraulic gradient determined by the infiltrometers and tensiometers, respectively. Darcy's Law for saturated water flow and the Green-Ampt Approximation for infiltration of water into a soil are used to calculate the conductivity of the liner. Darcy's Law is written as

$$Q=-k_{sat}IA \qquad [1]$$

where Q is discharge rate, k_{sat} is saturated hydraulic conductivity, I is the hydraulic gradient, and A is the cross-sectional area of flow. Q/A is the "average" steady-state infiltration flux from the ring infiltrometers and the water balance (Table 2). The gradient measured by the tensiometers at the end of both the first and second year of ponding is 2. The resultant calculated hydraulic conductivity ranges from 2.5×10^{-9} cm/sec to 4.0×10^{-8} cm/sec depending on the infiltration data used in the calculation (Table 3).

Table 3. Mean saturated hydraulic conductivity determined from infiltration fluxes using Darcy's Law and Green-Ampt Approximations.

Infiltrometers	Darcy's Law K_{sat} (cm/s)	Green and Ampt K_{sat} (cm/s)
Large Rings first year	2.5×10^{-9}	2.4×10^{-9}
Large Rings second year	2.9×10^{-9}	3.2×10^{-9}
Small Rings first year	4.0×10^{-8}	3.9×10^{-8}
Small Rings second year	3.0×10^{-8}	3.3×10^{-8}
Water Balance	3.2×10^{-8}	3.5×10^{-8}

Green and Ampt (1911) introduced the earliest equation for soil infiltrability which is still widely used. Their method assumes that the wetting front is sharp, the metric potential (tension) at the front is constant, and the wetted zone is uniformly saturated and of constant hydraulic conductivity. This approximation differs from Darcy's Law in that knowledge of the depth of the wetting front is required instead of hydraulic gradient. Under these assumptions, the analytical solution to vertical infiltration is:

$$k = i \left[1 + \frac{h + \psi_f}{L_f}\right] \qquad [2]$$

The bracketed term is the hydraulic gradient, h is the ponding depth, ψ_f is the metric potential at the letting front, L_f is the depth to the wetting front, and i is the steady-state infiltration flux.

The depth of the pond (h) was 29.5 cm for the first 406 days of ponding, and 31.0 cm thereafter; a weighted average of 30.2 cm is used for the pond depth to calculate the hydraulic conductivity from the second year of data. The tensiometer and water-balance data suggested that the wetting front had moved to a depth of approximately 35 cm the first year and to approximately 44 cm at the end of the second year. The metric potential at the wetting front is assumed to be the same as that of the tensiometers immediately below the wetting front, resulting in tension values of 7 and 6 cm of water for the first and second years, respectively. The hydraulic conductivity values calculated using the Green and Ampt assumptions are given in Table 3. All calculated hydraulic conductivity values are less than 5 x 10^{-8} cm/sec, and are thus, at a minimum, a factor of two less than the U.S. EPA standard for soil liners.

Conclusions

The monitoring of the field-scale liner began April 12, 1988, with the filling of the pond on the liner. Numerous difficulties have been encountered in data collection. Part of this problem is a direct result of the large volume of data from the instruments, which monitored on daily to monthly frequencies. However, no less significant is the very small measured fluxes. Minor perturbation in instrument output or ambient conditions around the liner may result in relatively large deviations in the data. For example, in calculating the overall infiltration volume from the pond into the liner, evaporation accounts for approximately 80% of the water loss from the pond; thus, variations in evaporation in the liner shelter and the ability to precisely measure evaporation rates have a major impact on the estimated volume of water infiltrating into the liner.

Based on current data and observations, monitoring of the field-scale liner will have to continue at least two more years to observe water break through. It is, therefore, premature to draw many conclusions or make recommendations. However, there are some implications of the results which can be drawn at this time from the results to date.

Construction

Our experience on this project, a review of the literature, and observation of commercial liners has lead us to the following conclusions:

- Standard engineering geology practices are adequate for sampling and selection of borrow materials used in construction of soil liners.
- Strict specifications of soil properties must be made in concert with rigid quality control to assure that a soil liner will be constructed and perform to design criteria. Specifications for an acceptable soil for use in a liner must include not only a maximum laboratory and/or *in situ* hydraulic conductivity, but should also include moisture content at time of compaction, maximum clod size, and minimum density and plasticity requirements.
- Soil moisture contents should be 1 to 3% wet of optimum as determined by a Standard Proctor Test.
- Liner materials should be processed before liner construction to insure a uniform moisture content, to reduce clod sizes to less than 5 cm, and to remove of all stones greater than 5 cm, and as many smaller stones as practical.
- The construction equipment must be large enough to fully compact the entire thickness of the lifts and have compactor feet which are at least as long as the compacted thickness of each lift, and preferable as long as the loose lift. Compaction should continue on each lift until a prescribed minimum density is measured at a reasonable number of locations before proceeding to the next lift.
- There must be bonding between lifts, and if the liner is to be constructed in phases.

Fluid Movement

Transport rates through the liner are effected by the physical state of the liner. The tensiometer results suggest that there is entrapped air in the field-scale liner, presenting some interesting secondary questions. When two fluids occupy a pore volume the effective permeability of the soil to each is decreased, in some cases to zero. Thus, the permeability to air may be zero, not allowing the air to escape while still reducing the effective permeability to water.

Liner Testing

Measurement of *in situ* saturated hydraulic conductivity is necessary to determine the adequacy of the liner. Ideally, hydraulic conductivity tests should be done on each lift after completion. However, a minimum of one conductivity test should be conducted on a small test pad constructed using the same QA/QC and design specifications as the full-scale liner. Testing the saturated hydraulic conductivity using double-ring infiltrometers is the most practical field test. Field infiltrometer tests should be conducted until a steady infiltration flux is achieved so that the actual conductivity can be calculated, rather than just long enough to determine that the saturated hydraulic conductivity is less than the maximum value. It should be noted that many of the instruments that were grouted in place leaked.

As a footnote to liner testing, we question the practicality of the practice of putting lysimeters under a liner to monitor leakage through the liner. Research on multi-layered covers for landfill sites has shown that the soil-gravel interface is a barrier to unsaturated flow (e.g. Cartwright, et al., 1989). The lysimeters vary in size; most vary from 3m x 3m to 6m x 6m, although larger ones are in use. The lysimeters are excavated below liner grade with a plastic membrane over the native material. The membrane is then covered by 40 to 60 cm of free-draining gravel, and a geofabric placed over the gravel to prevent liner soil from clogging the gravel. Depending upon the nature of the native soils and size of the lysimeter, downward moving water may tend to move laterally around the lysimeter to the native soil. Thus the breakthrough time and measured flux through the liner may be inaccurate, and only a major failure of the liner detected.

Acknowledgments

The authors wish to acknowledge the partial support of this study by the U.S. Environmental Protection Agency, Risk Reduction Engineering Laboratory, Cincinnati, Ohio, (project officer Dr. Michael H. Roulier) and the Illinois Hazardous Waste and Information Center (project officer Dr. Gary Miller). We appreciate the efforts of Drs. David Daniel and Andrew Rogowski for their guidance in the instrument installation and operation procedures. The generous donations of materials and equipment by Waste Management of North America, American Colloid Company, Gundle Lining Systems Inc., and, Clem Environmental Corporation are also recognized. We also thank the staff of the Illinois State Geological Survey for their efforts during the construction phases of this project, as well as the efforts of Kathy Cooley, Madelene Teirney, and Keith Mitchell for their technical support.

References

Albrecht, K. A., and K. Cartwright. 1989. Infiltration and hydraulic conductivity of a compacted earthen liner. Ground Water, v. 27, no. 1, pp. 14-19.

Albrecht, K. A., B. L. Herzog, L. R. Follmer, I. G. Krapac, R. A. Griffin, and K. Cartwright. 1989. Excavation of an instrumented earthen liner: Inspection of dyed flow paths and soil morphology. Hazardous Waste and Hazardous Materials, v. 5, no. 4, pp. 269-279.

Acar, Y. B., and S. D. Field. 1983. Organic leachate effects on the structural stability and hydraulic conductivity of clay liners, a proposal. Louisiana State University, Baton Rouge, LA, 11 p.

Anderson, J. L., and J. Bouma. 1973. Relationships between saturated hydraulic conductivity and morphometric data of an argillic horizon. Soil Science Society America Proceedings, v. 37, pp. 408-413.

Cartwright, K., T. H. Larson, B. L. Herzog, T. M. Johnson, K. A. Albrecht, D. L. Moffett, D. A. Keefer, and C. J. Stohr. 1988. Trench covers to minimize infiltration at waste disposal sites. Illinois State Geological Survey Circular 541, 88 p.

Chen, H. W., and L. O. Yamamoto. 1987. Permeability Tests for Hazardous Waste Management Unit Clay Liners. In D.J.A. van Zyl, S. R. Abt, J. D. Nelson, and T. A. Sheppard (eds.), Geotechnical and Geohydrological Aspects of Waste Management, Lewis Publishers, Chelsea, MI, pp. 229-244.

Daniel, D. E. 1981. Problems in predicting the permeability of compacted clay liners. Symposium on Uranium Mill Tailings Management, Fort Collins, CO, p. 665-675.

Daniel, D. E. 1984. Predicting hydraulic conductivity of clay liners. Journal of Geotechnical Engineering, v. 110, no. 2, pp. 285-300.

Daniel, D. E., D. C. Anderson, and S. S. Boynton. 1984. Fixed-wall vs. flexible-wall permeameters. In Hydraulic Barriers in Soil and Rock, A. I. Johnson, et al. (eds.), ASTM Special Technical Publication 874, pp. 107-126.

Daniel, D. E., S. J. Trautwein, S. S. Boynton, and D. E. Foreman. 1984. Permeability testing with flexible-wall permeameters. Geotechnical Testing Journal, v. 7, no. 3, pp. 113-120.

Daniel, D. E., and S. J. Trautwein. 1986. Field Permeability Test for Earthen Liners. In S. P. Clemence (ed.), Use of *In Situ* Tests in Geotechnical Engineering. American Society of Civil Engineers, New York, NY, pp. 146-160.

Daniel, D. E., and K. W. Brown. 1988. Landfill Liners: How Well do they Work and What is Their Future? In Gronow, Schofield, and Jain (Eds.), Land Disposal of Hazardous Water, Ellis Horwood Limited, West Essex, England, pp. 235-244.

Dunn, R. J., and J. K. Mitchell. 1984. Fluid conductivity testing of fine-grained soils. Teotechnical Engineering, v. 110, no. 11, pp. 1648-1665.

Elsbury, B. R., G. A. Sraders, D. C. Anderson, J. A. Rehage, J. O. Sai, and J.D.E. Daniels. 1988. Field and Laboratory Testing of a Compacted Soil Liner. Draft Final Report to U.S. EPA, Cincinnati, OH. Contract No. 68-03-3250.

Ely, R. L., G. L. Kingsbury, M. R. Branscome, L. J. Goldman, C. M. Northeim, J. H. Turner, and F. D. Mixon. 1983. Performance of clay caps and liners for disposal facilities (draft). Grant 68-03-3149 to Research Triangle Institute, U.S. EPA, Cincinnati, OH.

Foreman, D. E. 1984. The effects of hydraulic gradient and concentrated organic chemicals on the hydraulic conductivity of compacted clay. Unpublished M.S. thesis, University of Texas at Austin, 346 p.

Green, W. H., and G. A. Ampt. 1911. Studies on soil physics: I. Flow of air and water through soils. Journal of Agricultural Science, v. 4, no. 1, pp. 1-24.

Green, R. E., and J. C. Corey. 1971. Calculation of hydraulic conductivity: A further evaluation of some predictive methods. Soil Science Society of America Proceedings, v. 35, p.p. 3-8.

Grim, R. E., 1962. Applied Clay Mineralogy. McGraw-Hill, NY, 422 p.

Grim, R. E. 1968. Clay Mineralogy. McGraw-Hill, NY, 596 p.

Herzog, B. L., and W. J. Morse. 1986. Hydraulic conductivity at a hazardous waste disposal side: Comparison of laboratory and field-determined values. Waste Management and Research, v. 4, no. 2, pp. 177-187.

Haynes, D. B. 1990. Temperature variations effect on field-measured infiltration. Soil Science Society of America Journal, v. 54, pp. 305-312.

Johnson, A. I. 1954. Symposium on soil permeability, ASTM STP 163. American Society for Testing and Materials, Philadelphia, PA, pp. 98-114.

Johnson, T. M., T. H. Larson, B. L. Herzog, K. Cartwright, C. J. Stohr, and S. J. Klein. 1983. A study of trench covers to minimize infiltration at waste disposal sites: Task II Report, Laboratory Evaluation and Computer Modeling of Trench Cover Design, NUREG/CR-2478, v. 2, 94 p.

Krapac, I. G., K. Cartwright, K. Albrecht, D. F. Brutcher, P. B. DuMontelle, L. R. Follmer, R. A. Griffin, B. R. Hensel, B. L. Herzog, T. H. Larson, K. W. Miller, S. V. Panno, K. H. Rehfeldt, J. B. Risatti, and W. Su. 1989. Field study of transit time through compacted clays. Draft Final Report. U.S. EPA, Cincinnati, OH, 241 p.

Lambe, T. W. 1958. The structure of compacted clay. Journal of the Soil Mechanics Foundation Division, American Society of Civil Engineering, v. 84, no. SM2, p. 34 (paper 1654).

Mason, D. D., J. F. Lutz, and R. G. Peterson. 1957. Hydraulic conductivity as related to certain soil properties in a number of great soil groups--sampling errors involved. Soil Science Society of America Proceedings, v. 21, pp. 554-561.

Mitchell, J. K. 1976. Fundamentals of soil behavior. John Wiley and Sons, Inc., NY, 422 p.

Mitchell, J. K., D. R. Hooper, and R. G. Campanella. 1965. Permeability of compacted clay. Journal of the Soil Mechanics Foundation Division, American Society of Civil Engineers, v. 91, no. SM4, pp. 41-65.

Mitchell, J. K., and J. S. Younger. 1967. Abnormalities in hydraulic flow through fine-grained soil. Standard Technical publication 417, ASTM, Philadelphia, PA, pp. 106-139.

Nielson, D. R., J. W. Biggar, and K. R. Ehr. 1973. Spatial variability in field-measured soil-water properties. Hilgardia, v. 42, pp. 215-259.

Olson, R. E., and D. E. Daniel. 1979. Field and laboratory measurements of the permeability of saturated and partially saturated fine-grained soils. ASTM Symposium on Permeability and Groundwater Contaminant Transport, Philadelphia, PA, 67 p.

Panno, S. V., B. L. Herzog, K. Cartwright, K. R. Rehfeldt, I. G. Krapac, and B. R. Hensel. 1990. Infiltration of Water into a Field-Scale Compacted Soil Liner. Ground Water, in press.

Parkin, T. B., J. J. Meisinger, S. T. Chester, J. L. Starr, and J. A. Robinson. 1988. Evaluation of statistical estimation methods for lognormally distributed variables. Soil Science Society of America Journal, v. 52, pp. 323-329.

Rogowski, A. S. 1972. Watershed Physics: Soil Variability Criteria, Water Resources Research, vol. 9, pp. 1015-1023.

Rogowski, A. S. 1989. Relationship of Laboratory and Field Determined Hydraulic Conductivity in Compacted Layer. U.S. Department of Agriculture Northeast Watershed Research Center, University Park, PA. Draft Final Report of U.S. EPA, Cincinnati, OH, 247 p.

Seed, H. B., and C. K. Chan. 1959. Structure and strength characteristics of compacted clays. Journal of the Soil Mechanics and Foundation Division, American Society of Civil Engineers, v. 85, no. SM5, pp. 87-128.

Smith, R. M., and D. R. Browning. 1942. Persistent water-unsaturation of natural soil in relation to various soil and plant factors. Soil Science Society of America Proceedings, v. 7, pp. 114-119.

Su, W. J., P. B. DuMontelle, D. F. Brutcher, and K. Cartwright. Methodology and guidelines for evaluating soil as earthen liner material. In preparation.

U.S. Environmental Protection Agency. 1979. Design and construction of covers for solid waste landfills. Municipal Environmental Research Laboratory, Cincinnati, OH, EPA-600/2-79-165, 249 p.

U.S. Environmental Protection Agency. 1983. Lining of waste impoundment and disposal facilities. U.S. Government Printing Office, 1983 0-400-647, 448 p.

U.S. Environmental Protection Agency. 1988. Design, Construction, and Evaluation of Clay Liners for Waste Management Facilities. U.S. EPA Risk Reduction Engineering Laboratory, Cincinnati, OH EPA/530-SW-86-007F.

U.S. Federal Register. 1987. Hazardous Waste Management System: Minimum Technology Requirements. pp. B1-B10, April 17, 1987.

van Olphen, H. 1963. An introduction to clay colloid chemistry. Inter-science, John Wiley and Sons, NY, 301 p.

Young, R. N., and B. P. Warkentin. 1975. Soil properties and behavior. Geotechnical Engineering 5. Elsevier Scientific Publishing Company, NY, 449 p.

Zimmie, T. F., J. S. Doynow, and J. T. Wardell. 1981. Permeability testing of soils for hazardous waste disposal sites. Proceedings of the Tenth International Conference on Soil Mechanics and Foundation Engineering, v. 2, Stockholm, Sweden, pp. 403-406.

Detailed Case History of Clay Liner Performance

Denys W. Reades[1], Leo R. Lahti[1], Robert M. Quigley[2]
and Angelos Bacopoulos[3]

Abstract

The Keele Valley Landfill at Maple, Ontario, consists of a 99 ha (245 acres) facility designed to serve the Metropolitan Toronto area. The landfill is located in a former sand and gravel pit, with the base and the sides of the pit lined with at least 1.2 m of low plasticity clay till compacted to achieve the required hydraulic conductivity equivalent to 1×10^{-8} cm/s or less.

This paper describes the results of a comprehensive program of monitoring the in situ performance of the clay till liner since construction was initiated in 1983. Results of monitoring to date indicate that the liner is performing as required and good agreement between the results of laboratory hydraulic conductivity tests and the in-situ liner performance monitoring program have been obtained. Based on results to date, the rate of contaminant migration into the liner is governed by diffusion.

Introduction

The Keele Valley Landfill at Maple, Ontario, approximately 10 km north of Metropolitan Toronto (Figure 1), is owned and operated by the Municipality of Metropolitan Toronto Department of Works and is used for the disposal of municipal solid waste. It is located in an existing 99 ha sand and gravel pit that ranges up to 30 m in depth and has a capacity for about 20 million tonnes of refuse. The subgrade beneath the base of the pit consists of sand, with the phreatic surface approximately 1 m to 3 m below the base of the pit. Two overburden

---

[1] Principal, Golder Associates Ltd., Mississauga, Ontario.

[2] Professor, The University of Western Ontario, London, Ontario

[3] Assistant Director, Solid Waste Management Division, Metropolitan Toronto Department of Works, Toronto, Ontario

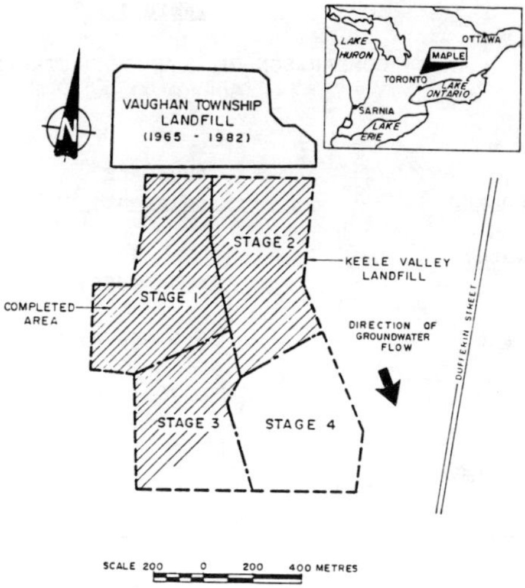

Figure 1 Site Location Plan

aquifers are separated by an intermediate layer of low permeability. In both the upper and lower aquifers, groundwater flow is to the southeast.

Construction of the low permeability 1×10^{-8} cm/s landfill liner began in 1983. The landfill is divided into four approximately equal areas (Stages 1 to 4) (Figure 1) and the liner for Stages 1 & 2 has been completed. Stage 3 is now nearly completed. The total area of clay till liner placed to date is approximately 73.5 ha (181 acres). A leachate collection system is installed on top of the liner to minimize hydraulic head on the liner.

Acceptance of the clay liner to meet the criterion of 1×10^{-8} cm/s is based on both the results of laboratory permeability tests on undisturbed Shelby tube samples of the clay liner extracted and tested during each construction season, and the results of the extensive field program of liner performance monitoring. This paper details the results of the in situ liner performance monitoring program to date and compares them with the results from the laboratory testing program.

The clay till used until mid-1989 was obtained from two borrow areas immediately adjacent to the landfill site, with the major amount obtained from the Avondale borrow pit. Since 1989, clay for the liner has been brought in from an outside borrow pit some 27 km away (the Jenkins borrow pit). The properties of the clay from each borrow pit are summarized on Table 1.

Table 1

COMPARISON OF CLAY CHARACTERISTICS
BORROW AREAS

TEST PERFORMED	AVONDALE (adjacent)	JENKINS (off-site)
In situ water content %	14.3 (666)	14.6 (109)
Clay content %	25.0 (674)	31.1 (134)
Specific gravity	2.70 (20)	2.68 (30)
Atterberg limits		
Liquid limit	25.1 (132)	29.1 (38)
Plastic limit	14.7 (132)	16.7 (38)
Plasticity Index	10.4 (132)	12.4 (38)
Standard Proctor		
Max. density, pcf	121.4 (57)	117.1 (17)
Optimum water, %	12.3 (57)	14.1 (17)
Areal mean permeability cm/s ($\sigma_3' = 165$ kPa)	7.87×10^{-9} (180)	8.56×10^{-9} (41)

Number of tests indicated in brackets.

Trial Liner Sections

Three trial liner sections were constructed in 1980 and 1981 in advance of the main liner construction with various equipment and using two types of clay at varying water contents (see Lahti et al, 1987). The main conclusions of the trial liner sections were as follows:

1. The 30,000 kg compactor, with deeper penetrating tampers (feet), was more effective in compacting the 150 mm thick lifts of the till than was a lighter 15,000 kg compactor with shallower tampers;

2. Hydraulic conductivity tests on both Shelby tube and block samples indicated that, for samples compacted above Standard Proctor optimum water content, the hydraulic conductivity of the till appeared to be dependent primarily upon the percentage of clay particles present;

3. For clay contents in excess of 20 per cent by weight (Till 'A'), the hydraulic conductivity determined from 7 laboratory permeability tests on clay samples extracted from the trial liner sections ranged from 5 x 10^{-9} cm/s to 1 x 10^{-8} cm/s.

Based on the experience gained from the trial liner sections, it was concluded that the required in situ hydraulic conductivity of less than 1 x 10^{-8} cm/s could be achieved using the clayey till from the adjacent borrow areas, provided that the clay content was greater than 20 percent and that the till was compacted at least 95 percent of the Standard Proctor maximum dry density at a water content 2 to 3 percent above the optimum water content.

Leachate-Clay Liner Compatibility Tests

The initial leachate-clay liner compatibility program was carried out between 1984 and 1986, primarily by Professor Quigley at the University of Western Ontario.

Mineralogical characterization of the clay till borrow materials initially focused on the difference between the upper brown weathered clay and lower grey unweathered clay in the borrow pit. The upper brown clay was composed of about 38 percent quartz and feldspar, 7 percent borderline smectite/vermiculite and more chlorite (approximately 8 percent).

The potential influence of the smectite/vermiculite content of the brown clay and behaviour of the clay when exposed to leachate was considered significant, due to the potential for chemical interaction with leachate by C-axis contraction of vermiculite and expansion or contraction of the double layers of adsorbed cations and water. However, hydraulic conductivity tests using Keele Valley leachate as the permeant under a confining pressure of 160 kPa indicated no change in hydraulic conductivity even after the passage of 12 pore volumes of leachate. At gradients of 240 to 300, this represented many centuries of advection in the field where a much lower gradient exists (less than 3). Although significant mineralogical changes occurred, these did not influence the measured hydraulic conductivity of the samples because they consolidated simultaneously with the chemical interactions.

Diffusion testing on samples of the clay liner at the University of Waterloo during this time period indicated an effective diffusion coefficient for chloride of between 7.0 and 7.2 x 10^{-6} cm/s.

Construction Quality Control

Construction of the liner has been carried out by the Municipality of Metropolitan Toronto Works Department, with quality control for the liner construction and performance monitoring provided by Golder Associates. In addition, Trow Ltd. has been retained by the Ministry of the Environment as an independent third party to carry out check testing and to provide an overall review role with respect to construction of the liner.

Clay till for the base liner was spread in 200 mm to 300 mm thick loose lifts with scrapers, then bladed by bulldozers and compactors to 100 mm to 150 mm average lift thickness. Cobbles and boulders larger than 150 mm size were removed from the incoming liner material. Each lift of the liner was compacted with 30,000 kg padfoot type compactors to a minimum of 95 per cent of the Standard Proctor maximum dry density. The slope liner was placed and compacted in successive 5 m wide strips in a similar manner as the base liner, but using standard 15,000 kg gross mass compactors; using heavy 30,000 kg compactors on the slopes led to excessive bulging and lateral deformation.

The incoming Till 'A' liner material generally had a natural water content ranging from 2 per cent below the optimum water content to 3 per cent above optimum. Due to rapid evaporation on hot sunny days, the liner materials had to be watered continually by two water trucks. Each completed lift was wetted, and once a loose lift had been spread, additional water was added prior to blading and compaction.

During dry weather, liner material placed during the previous day was watered at the start of the day and that portion of the liner which had been completed was watered at the end of each shift. Shortly after a section or strip of base liner was completed, it was covered with a 300 mm thick protective sand layer to prevent the liner surface from drying and cracking. This protective cover was left on permanently and watered frequently during dry weather periods.

The field testing consisted principally of in situ density and natural water content testing of the sand subgrade and clay liner material using a Troxler Model 3411B nuclear density apparatus. It was found essential to plot density test results daily on a dry density vs. water content graph to verify that the till was being placed within an acceptable combination of a suitable density and water content (2 to 3 per cent wet of optimum water content) so that a high degree of saturation was being maintained (approximately 95 per cent after compaction). A typical plot of dry density vs. water content for a one-week period is given in Figure 2.

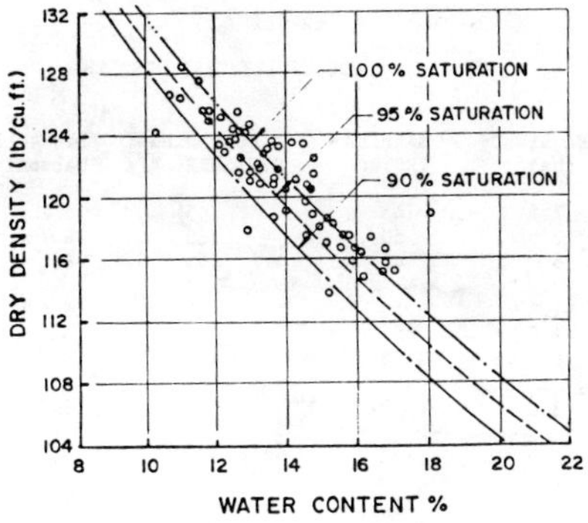

Figure 2 Typical plot of in situ dry density vs. water content for clay liner.

Over the seven-year period to date (1983-1989), approximately 1,180,000 m^3 of liner material has been placed over an area of about 73.5 ha, of which 57.7 ha are base liner. During the course of the construction, 4,581 in situ density tests have been performed on the liner, and 105 Proctor compaction tests, and 861 hydrometer analyses conducted in the field laboratory. These tests have provided field verification that the criteria established with respect to density and clay content of the liner are being met. However, final approval of the in-place liner is based on the results of laboratory hydraulic conductivity tests.

Hydraulic Conductivity Tests

A comprehensive hydraulic conductivity testing program has been carried out each year on thin-walled Shelby tubes samples extracted from the clay till liner as it is placed. As described in Lahti et al (1987), a detailed program was initially set up to examine the influence of sample size, depth of advance, hydraulic gradient and cell pressure. A protocol for hydraulic conductivity testing of the liner each year was also developed (see Richards and Thompson, 1989). This requires that one test is carried out for about every 3,000 m^3 of compacted liner and that the results for each year are analyzed statistically to ensure compliance with the 1×10^{-8} cm/s hydraulic conductivity criterion for the liner.

Table 2

HYDRAULIC CONDUCTIVITY TEST RESULTS

YEAR	STAGE	AREA LINED (ha)	SAMPLES TESTED	CONFINING PRESSURE, kPa	HYDRAULIC CONDUCTIVITY, cm/s Harmonic Avg.	Geometric Mean
1983	1	7.5	45	165	6.2×10^{-9}	8.1×10^{-9}
1984	1	12.5	50	165	6.7×10^{-9}	8.2×10^{-9}
1985	1	8.0	45	165	6.1×10^{-9}	7.5×10^{-9}
1986	2	11.8	50	165	6.6×10^{-9}	7.9×10^{-9}
1987	2	7.0	52	165	6.3×10^{-9}	7.2×10^{-9}
1988	2	6.2	28	165	6.2×10^{-9}	7.2×10^{-9}
1988	3	3.0	9	145	9.2×10^{-9}	1.06×10^{-9}
1989	3	15.0	70	145	7.0×10^{-9}*	7.7×10^{-9}*

* Liner thickened: results are for equivalent design thickness of 1.2 m

The overall hydraulic conductivity values determined for each of the past years are given in Table 2. Tests within any given Stage of the landfill were conducted under a confining pressure corresponding to half the average height of refuse within that stage. This was originally judged to be a reasonable conservative assessment of a confining pressure which would be expected to be exerted on the base of the landfill within a period of 2 to 4 years. However, due to an increased rate of filling, this height of refuse has often been reached in 1 to 2 years and the refuse has approached full height (up to 40 m height) generally within 3 years of placement of the base liner in any area.

Laboratory hydraulic conductivity tests for Stages 1 and 2 were conducted at a confining pressure of 165 kPa, while a reduced confining pressure of 145 kPa was used for Stage 3 where the refuse thickness was less. Although the influence of confining pressure on hydraulic conductivity had been recognized in 1983 at the start of the project, the comparatively small reduction in confining pressure from 165 kPa to 145 kPa had a much greater impact than anticipated on the hydraulic conductivity test results (see Table 2).

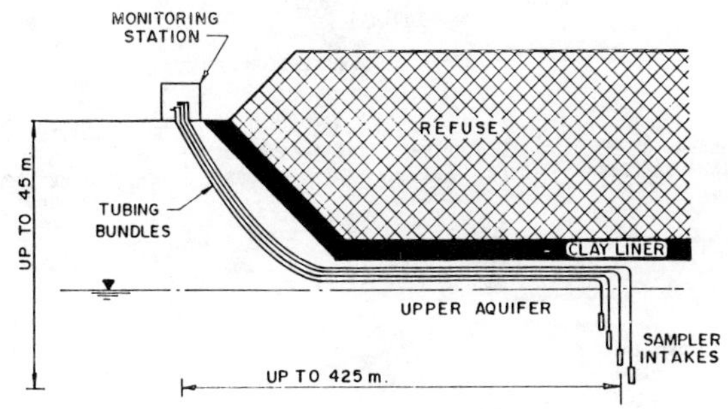

Figure 3 Schematic section showing sub-liner groundwater samplers.

It should also be noted that the liner monitoring program is only one component of a number of on-site monitoring and evaluation programs concerning the liner. Samples of liner exposed to leachate have been cored to determine the depth of actual solute migration into the liner. As well, groundwater beneath and around the site is monitored on a regular basis.

Instrumentation Overview to the End of 1989

One hundred and ten sub-liner water samplers have been installed to date in the upper aquifer below the landfill. The primary purpose of these samplers is to detect any significant leakage or catastrophic failure of the liner. Because of the uncertainty in sealing monitoring wells through the clay liner and operational difficulties associated with bringing vertical groundwater sampling installations up through an active landfill, it was decided to place the outlet tubes horizontally across the base of the pit and up the side slopes before the liner was placed (see Figure 3).

At each of the six initial nest locations, samplers were installed at 1.2 m intervals from just below the water table to depths of about 15 m into the aquifer beneath the landfill. This detailed vertical control was necessary in order to distinguish between leachate-derived contaminants from potential future migration through the clay liner and contaminants in an existing plume from the former Vaughan landfill north of the present site.

Table 3

SUMMARY OF INSTRUMENTATION

STAGE	NUMBER OF SUBLINER SAMPLERS	LYSIMETER DEEP	LYSIMETER SHALLOW	CONDUCTIVITY SENSORS	VIBRATING WIRE PIEZOMETERS
1	60	3	3	4	12
2	25	1	2	2	16
3	24*	2*	3*	3*	28*

* Installed or proposed

Although the majority of the base liner will be subjected to 30 to 40 m of refuse which will impose a pressure greater than that used for the laboratory tests, the thickness of the base liner will be increased from the original design thickness of 1.2 m to 1.68 m in areas of the landfill near the south end with less than about 20 m of refuse, in order to produce the equivalent seepage rate. A similar increase was also necessary in part of Stage 3 after higher than expected hydraulic conductivity results were obtained from a new trial liner using the off-site clay. Later results showed that the off-site clay was similar in quality to the clay from the adjacent borrow areas, but the thickened liner in Stage 3 assisted in compensating for the lower testing confining pressure in Stage 3 and higher hydraulic conductivity results where a surficial portion of the clay liner became frozen.

Preliminary tests to measure the degree of anisotropic permeability of the clay suggest that k_H is on average about 1.5 times k_v.

The laboratory hydraulic conductivity tests indicate that the quality of the clay liner placed to date in Stages 1, 2 and 3 has been satisfactory although this was only achieved in Stage 3 by the thickening of the liner. Nevertheless, the overall in situ performance of the clay liner is assessed by the comprehensive liner performance monitoring program described in the following section.

Liner Performance Monitoring Program

The liner performance monitoring program at the Keele Valley Landfill has been designed to assess the in situ performance of the clay liner by monitoring seepage rates and determining rates of leachate migration into the liner. As shown in Table 3, four types of devices have been or are being installed in Stages 1, 2 and 3 to monitor the performance of the liner: sub-liner water samplers, vibrating wire piezometers above and below the liner, areal collection lysimeters within and below the liner and conductivity sensors within the liner.

Because of the long distance (up to 425 m) and elevation differences (up to 45 m) between the sampler intake locations and the margin of the landfill where the samples are recovered, the sub-liner groundwater samplers must consist of some type of positive displacement device. Two basic types of commercially available groundwater samplers were used in alternating vertical positions at four nest locations: a flexible bladder pump and a gas displacement sampler (see King et al, 1986).

Only seventeen samplers have ceased to function since installation. Thirteen have been bladder pump samplers and four have been gas displacement samplers. Four of the bladder pump samplers and all four displacement samplers are at locations where reduced infiltration due to the liner has caused the water table to drop below the sampler. The primary reasons for the apparent failure of the remaining nine bladder style pumps are believed to be the slow rate of recharge and the clogging of check valves by silt. On the other hand, the gas displacement samplers produce larger volumes of water in shorter periods of time and, since 1986, only gas displacement samplers have been used.

Forty-six <u>vibrating wire piezometers</u> have been installed to date. Eighteen of these have been placed on the liner surface to monitor the build-up of leachate levels above the liner and the effectiveness of the leachate collection system. The remainder have been installed in the aquifer below the liner to monitor the decrease in the level of the water table. All readings from the vibrating wire piezometers are corrected for changes in barometric pressure and temperature since installation.

Seepage volumes through the liner are measured at eleven 15 m x 15 m square <u>areal collection lysimeters</u>. Each lysimeter is composed of a basal HDPE membrane with overlying sand filter layers. Seepage moving down through the clay liner is collected by the lysimeter and drained by gravity through an outlet tube to the margin of the liner.

Five of these lysimeters have been installed below the liner and six have been installed within specially thickened sections of the liner but with only 0.3 m or 0.5 m of clay till above them. The lysimeters were installed approximately midway between the drains which form part of the leachate collection system above the liner and therefore at locations where the leachate head is at a maximum. The quality of effluent from the shallow lysimeters within the liner is determined on a regular basis as solute concentrations gradually increase. So far, no flow has been observed from the five deep lysimeters constructed below the liner.

Seven <u>conductivity sensor sets</u> have been installed to date, each adjacent to a shallow lysimeter. Each set of sensors is composed of 12 arrays of electrodes installed at 7.5 cm vertical intervals in the clay liner. Each array shows increasing conductivity as increasing concentrations of dissolved ions derived from the leachate migrate downward.

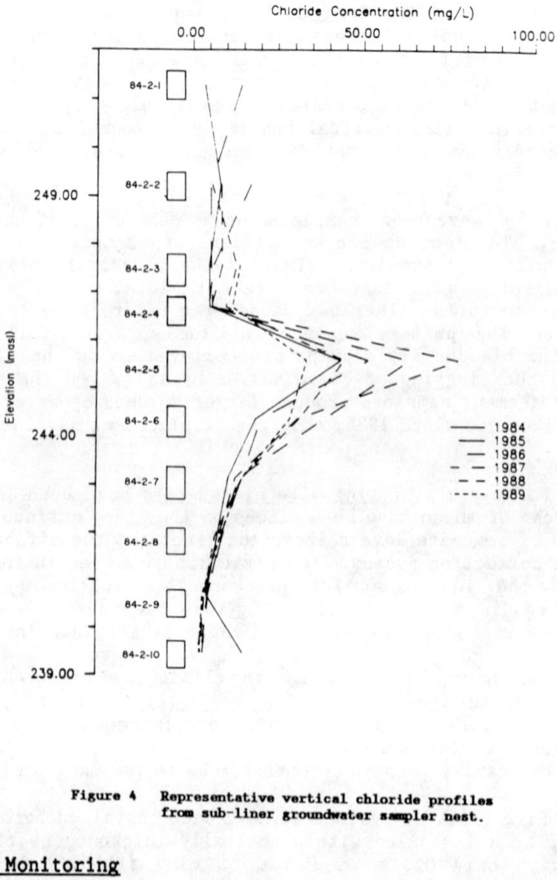

Figure 4 Representative vertical chloride profiles from sub-liner groundwater sampler nest.

Results of Monitoring

The sub-liner groundwater samplers have provided very good vertical profiles of the groundwater chemistry below the liner with time. Above background ion concentrations have been detected in many sampler nests. Generally, these zones of contamination have been observed since monitoring began and occur at relatively constant concentrations at elevations at which they can be attributed to one of a number of lobes of the contaminant plume emanating from the earlier un-lined Vaughan Landfill immediately upgradient of the Keele Valley site. At some locations, increases in ion concentrations have occurred since monitoring was initiated (see, for example, Figure 4). However, these have been vertically bounded and occurred at elevations at which contamination was expected to occur as the plume front advanced. The absence of elevated solute concentrations in the shallow samplers indicates that leachate solute from the Keele Valley site has not yet reached the underlying aquifer.

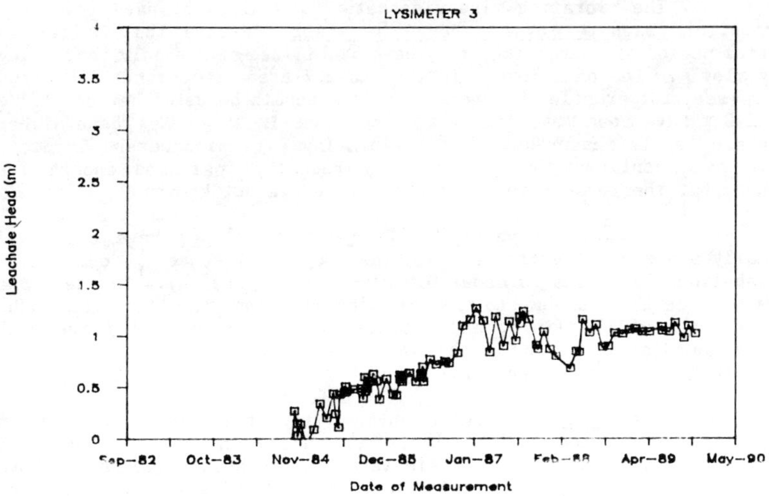

a) Above the liner

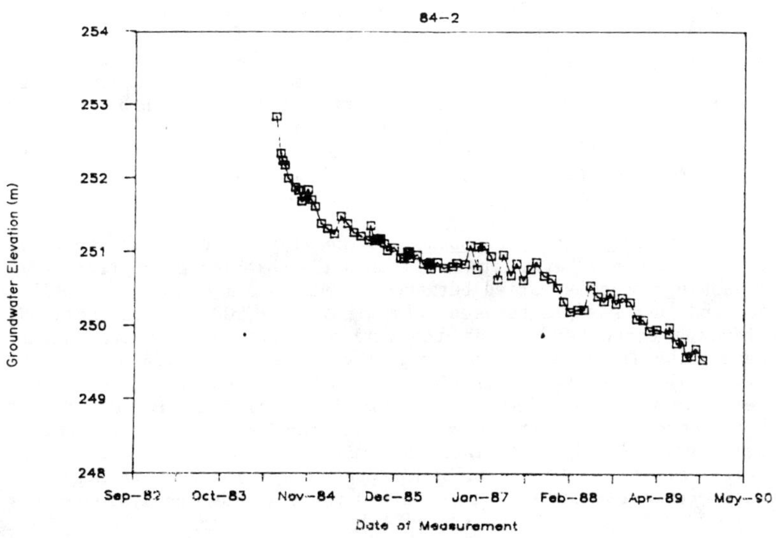

b) Below the liner

Figure 5 Water levels above and below the liner.

The vibrating wire piezometers installed at each lysimeter to monitor leachate levels above the liner indicate that, after an initial period of increasing, they have stabilized or are only increasing very slowly. Leachate levels in the Stage 2 area are between 1.0 m and 1.5 m (see, for example, Figure 5a) while leachate heads of between 1.7 m and 2.8 m have been measured at two locations in Stage 1. These higher leachate levels measured with the vibrating wire piezometers in Stage 1have been confirmed by 2 standpipes recently installed though the refuse, but the reasons for this difference are not known.

As expected water levels below the liner have decreased steadily since construction of the liner began. The average decline at 16 sub-liner locations is about 0.5 m/yr (see Figure 5b). The decrease in water levels is due to reduced infiltration resulting from the construction of the liner, and also to the operation of a purge well system immediately north of the Keele Valley Landfill to control the contaminant plume from the earlier Vaughan Landfill.

The in situ hydraulic conductivity of the thin 0.3 to 0.5 m clay liner above the shallow lysimeters has been assessed from the volumetric discharge from the lysimeters and the head of leachate above the liner. The average flow rates from the shallow lysimeters, and hence the hydraulic conductivity of the clay liner, have continued to decrease with time (see, for example, Figure 6). This reduction was not unexpected and was considered to be due to a combination of two possible factors. First, the clay has been subjected to increasing overburden pressure as the height of refuse is increased up to about 30 m. Second, bacterial clogging at the surface of the liner may be reducing the flow through the liner. Unfortunately, during 1989, flow from the two oldest shallow lysimeters installed in 1983 abruptly stopped. The reason for this is unknown. Attempts to blow nitrogen up on of the twin exit leads was unsuccessful in 1989 and further attempts to encourage flow will be made during 1990. At the end of 1989, the mean hydraulic conductivity of the thin layer of compacted clay till above the three shallow lysimeters for which data was complete fell within a range of 1.7×10^{-9} cm/s to 3.1×10^{-8} cm/s, with an average of about 9.0×10^{-9} cm/s. The hydraulic conductivity of the full 1.2 m thick liner should be equal or lower.

Solute concentrations in the lysimeter effluents have continued to increase with time. Although the chloride concentration in the leachate generally varied between 400 mg/L and 1,000 mg/L from 1984 to 1988 and increased to between 1,100 mg/L and 1,900 mg/L during 1989, the chloride concentrations in the effluents from the oldest shallow lysimeters installed in 1983 had only risen gradually by the end of 1988 to between 72 and 93 mg/L (see Figure 7) and only to about 54 mg/L in a lysimeter installed in 1984. Part of the delay in measuring "full" leachate concentrations is, however, due to the fact that leachate salts migrating down through the liner must mix with the original lysimeter porewater (chloride concentrations of about 13 mg/L) and the natural clay porewater (chloride concentrations of 50 mg/L) inside the lysimeter.

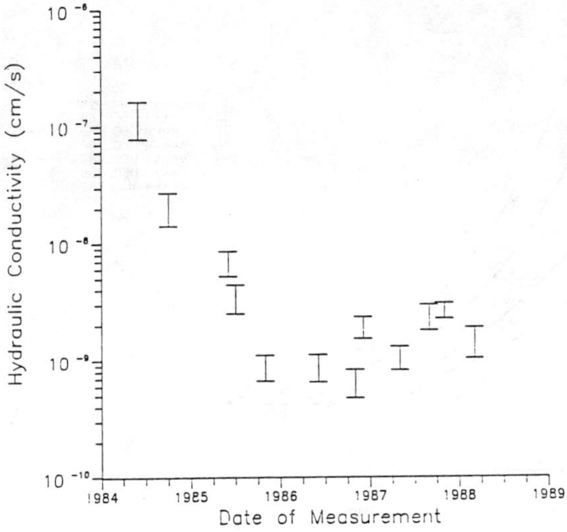

Figure 6 Hydraulic conductivity for thin clay till above a shallow lysimeter.

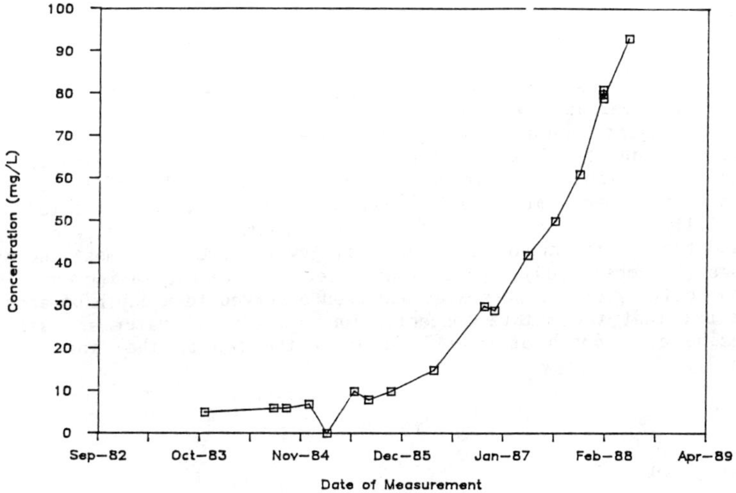

Figure 7 Effluent concentration from a shallow lysimeter.

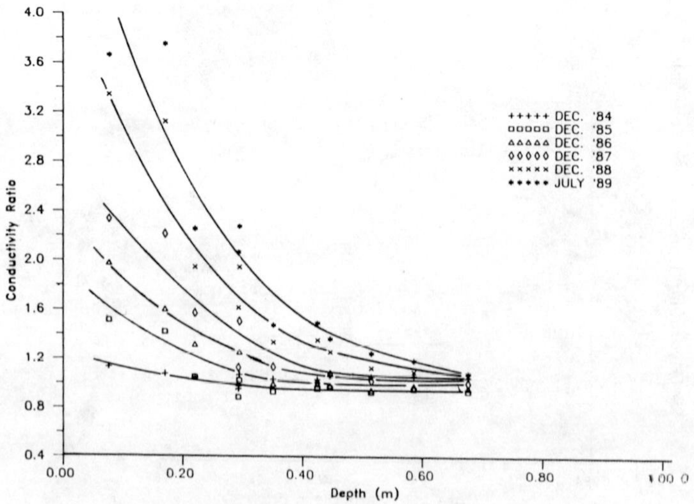

Figure 8 Conductivity ratio with time for a conductivity sensor set.

The results from the <u>seven conductivity sensor sets</u> are becoming less variable with time. Of the four installed in 1983 and 1984, all suggest some influence of leachate on the shallow arrays down to a depth of about 0.75 m. Initially there appeared to be considerable scatter in the results from three of these but during 1988 and 1989, the results have become similar with a much narrower spread when plotted as conductivity ratio (the ratio between the apparent conductivity at any time and the initial apparent conductivity) with time. The most recent readings (December 1989) suggest that after five years, an increase in the conductivity of the porewater has been observed to a depth of about 0.75 m but that the solute concentrations in the porewater are still increasing at a depth of only 7 cm below the top of the liner (see Figure 8).

The results from a conductivity sensor set installed in 1986 initially appeared to suggest that leachate had moved to a maximum depth of 0.7 m and was advancing through the liner at a faster rate at this location than elsewhere. However more recent readings indicate a levelling off of apparent conductivity values suggesting that leachate migration is slowing. This may be due to increased refuse thickness.

In summary, the liner performance monitoring instrumentation installed at Keele Valley indicates that:

- there has been no indication from the sub-liner samplers that salts in the leachate from the Keele Valley site have reached the underlying aquifer
- leachate heads above the liner are less than 3 m and except for Stage 1, less than 1.5 m.
- hydraulic conductivity for the shallow lysimeters fell within a range from 1.7×10^{-9} cm/s to 3.1×10^{-9} cm/s with an average of 9.5×10^{-9} cm/s at the end of 1988, however flow stopped from two of the oldest shallow lysimeters in 1989
- effluent concentrations of chloride in the shallow lysimeters have increased gradually with time but are still less than 100 mg/L even in those installed in 1983
- four of the conductivity sets show salt concentration increases down to a maximum depth of about 0.75 m after five years; the fifth was showing a more rapid advance but this slowed during 1989.

Diffusion Profiles

Two major liner exhumations have been carried out at the Keele Valley Landfill as part of the clay barrier performance monitoring program with Shelby tube samples extracted for subsequent chemical analyses. On each occasion, the upper 5 cm to 10 cm of the 30 cm thick sand protective layer on top of the clay liner was observed to be darkened by a black slime. In addition, the underlying 10$\pm$ cm of sand was observed to have been reduced to a grey colour from the original light brown colour.

A diffusion profile for chloride concentrations for a section of the 1984 liner which was excavated in December, 1988, is presented on Figure 9. The results show that the total depth of penetration of chloride after four years is about 70 cm; 31 cm through the sand and 40 cm into the clay.

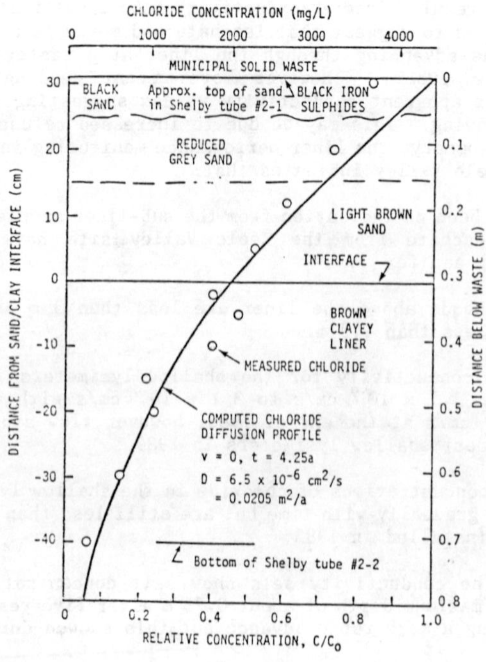

Figure 9 **Porewater chloride profile.**
Calculated and fitted chloride diffusion profile using D = 6.5 x 10^{-6} cm^2/s and v = 0.

The most interesting aspect of the diffusion curves is the continuity of the chloride profile through the sand/clay interface and a diffusion profile that starts at the top of the sand rather than at the top of the clay liner. This adds extra "effective thickness" to the liner system.

A computer chloride diffusion curve could be fitted to the field diffusion profile using a diffusion coefficient of 6.5 x 10^{-6} cm/s, which is very close to that of 7.0 x 10^{-6} cm^2/s obtained in earlier laboratory testing at the University of Waterloo.

Recent work also suggests that organic compounds such as toluene and ethylbenzene are also attenuated within a similar distance within a clay liner of similar age.

In previous work initiated in 1985, a one-dimensional advection-dispersion model was used to predict the performance of the liner at Keele Valley, using the analogy of a semi-infinite column with a source of constant concentration entering one end (one-dimensional Ogata Banks solution). Predictions were made using an average linear velocity of 2.9 cm/yr based on the results of the laboratory hydraulic conductivity testing, leachate heads measured in the landfill, a porosity of 0.3, an effective diffusion coefficient of 7 x 10^{-6} cm^2/s based on the testing at the University of Waterloo, and an assumed longitudinal dispersivity of 0.1.

Theoretical concentration profiles for a constant source were used to predict the chloride concentrations from the shallow lysimeters and the depth of migration of the leachate front. Each year, the depth of influence of leachate derived non-reactive solute, as detected by the conductivity sensors and shallow lysimeters, has been less than predicted by the model. However, there is good agreement between the conductivity sensor data and diffusion profile data for depth of leachate influence at the end of 1988 (0.5 and 0.4 meters from the top of the liner, respectively, after four years). This again suggests that diffusion governs leachate migration into the liner and, moreover, that the diffusion barrier begins at the top of the sand cushion above the liner rather than at the liner surface. Further data from the on-going laboratory and in situ monitoring of the site will continue to increase understanding of the long term performance of the liner.

Conclusions

Construction of Stages 1, 2 and 3 of the Keele Valley Landfill at Maple has followed rigorous quality assurance/quality control procedures during placement of the clay liner in addition to extensive field quality control testing. Laboratory hydraulic conductivity testing of the clay liner since 1983, indicates that the overall bulk hydraulic conductivity of the liner each year has been less than the specified value of 1 x 10^{-8} cm/s, although the liner was thickened in 1989.

Although additional data from future monitoring will increase understanding of liner performance, the results of the liner performance monitoring program to date indicate that the liner is performing better than a 1.2 m liner with a hydraulic conductivity less than 10^{-8} cm/s. This is based on the hydraulic conductivity of the thin section of clay liner above the shallow lysimeters and the results from the majority of conductivity sensor sets. Indeed, diffusion profiles from sections of the liner indicate that the rate of contaminant migration into the liner is governed by diffusion. No evidence has been found for hydraulic conductivities higher than those measured on undisturbed samples in the laboratory.

The good agreement between results of the laboratory testing and the in situ field measurements are very encouraging, and are considered to be due to proper construction techniques and quality control in the field.

Acknowledgements

The authors are indebted to the Municipality of Metropolitan Toronto, Department of Works, for permission to publish the information contained in the paper. The significant contributions of Metro Toronto, Golder Associates Ltd. and Trow Ltd. field, laboratory and office staffs in the liner construction and quality assurance are gratefully acknowledged.

Appendix 1 - References

King, K.S., S. Feenstra, D.W. Reades and A. Bacopoulos, (1986). Sub-Liner Groundwater Monitoring at a Large Lined Landfill. Presented at Sixth National Symposium and Exposition of Aquifer Restoration and Groundwater Monitoring, May 19-22, Columbus, Ohio.

Lahti, L.R., K.S. King, D.W. Reades & A. Bacopoulos (1987). Quality Assurance Monitoring of a Large Clay Liner. Proceedings, Geotechnical Practice for Waste Disposal, GT Division ASCE, Ann Arbor, Michigan, pp. 640-654.

Quigley, R.M., Yanful, E.K. and Fernandez, F. (1987). Ion Transfer by Diffusion Through Clayey Barriers. Proceedings of Geotechnical Practice for Waste Disposal '87, GT Division ASCE, Ann Arbor, Michigan, pp. 137-158.

Richards, P.A.L. and C.D. Thompson (1989). Permeability Protocol for the Compacted Clay Liner at Metropolitan Toronto's Keele Valley Landfill. Canadian Journal of Civil Engineering, Vol. 16, No. 4.

Reades, Denys W., K. Scott King, Eugene Benda, Robert M. Quigley, Keith Lesarge and Cheryl Heathwood (1989). The Results of On-Going Monitoring of the Performance of a Low Permeability Liner. National Well Water Association Conference, Kitchener, Ontario.

Summary Review of Construction Quality Control for Compacted Soil Liners

David E. Daniel[1], M. ASCE

Abstract

Construction quality control (QC) testing is crucial to the successful performance of compacted soil liners and covers. Construction QC is designed to verify that (1) materials used in construction are adequate; (2) the methods of construction are acceptable; and (3) liners and covers are adequately protected during and after construction. This paper presents a review of QC testing requirements and presents recommendations for required tests, sampling patterns, frequency of testing, handling of outlier points, and evaluation of systematic errors.

Introduction

Waste containment systems often include one or more hydraulic barriers constructed of earthen materials. Such barriers are called "soil liners," "compacted soil liners," "compacted clay liners," "natural material liners," and other similar terms. The compacted soil liners, whether used in a liner system located beneath buried waste or in a cover system placed over buried waste, must have a low hydraulic conductivity and other essential properties, e.g., adequate strength and ductility.

There are many facets of good design, construction, and testing of earthen liners. Examples of good performance of well constructed liners are described by Lahti et al. (1987), Gordon et al. (1989), and Johnson et al. (1990). Mitchell et al. (1965), Daniel (1987), Oakley (1987), Lahti et al. (1987), Herrmann and Elsbury (1987), Goldman et al. (1988), and Johnson et al. (1990) review compaction and construction methodology. Information on quality assurance for soil liners is provided by Gordon et

[1] Assoc. Prof. of Civil Engrg., Univ. of Texas, Austin, TX 78712.

al. (1984), U.S. EPA (1986), Lahti et al. (1987), Goldman et al. (1988), Boutwell and Hedges (1988), and Daniel and Benson (1990).

The purpose of this paper is to review quality assurance (QA) and quality control (QC) for compacted soil liners. The most important factors to be monitored are discussed, and testing procedures and frequencies are reviewed.

Achievement of Low Hydraulic Conductivity

The most important factors in achieving low hydraulic conductivity in compacted soil liners are:

1. Using suitable materials.
2. Placing the soil at the correct water content.
3. Properly preparing the surface to receive a lift of soil.
4. Compacting the soil with adequate passes from the proper type of compactor.
5. Protecting each compacted lift from damage.

Materials. Materials used to construct soil liners must contain adequate clay to achieve low hydraulic conductivity. Direct measurement of clay content is awkward and time consuming. It is the author's experience that soils can usually be compacted to achieve a hydraulic conductivity (k) $\leq 1 \times 10^{-7}$ cm/s if:

1. The amount of fines (defined as dry-weight percentage passing the No. 200 sieve) is ≥ 30%.

2. The plasticity index (PI) of the soil is ≥ 10%. For some materials, more stringent requirements may be appropriate, and in other cases, slightly lower PI's may prove to be adequate, e.g., Albrecht and Cartwright (1989), who measured an in situ k of 4×10^{-8} cm/s on a soil with a PI of 7%.

Stones or rocks can increase the hydraulic conductivity if too many are present or if the stones are too large. Although definitive data are lacking, the author recommends as a conservative guideline that:

1. The amount of gravel (defined as dry-weight percentage retained on the No. 4 sieve) be ≤ 10 to 20%.

2. The largest particle size be no more than 25 to 50 mm (1 to 2 inches).

Dry, hard clods of soil are difficult to compact and should be mechanically pulverized prior to adding water. If bentonite must be added to the soil, the soil/bentonite blend should be thoroughly mixed with mechanical tillers or mixed in a pug mill after the soil to be amended has been suitably pulverized.

Water Content. Low hydraulic conductivity cannot be achieved if the soil is too dry at the time it is compacted (Mitchell et al., 1965, and others). Conversely, the soil must not be too wet or it will be difficult to compact and will be susceptible to desiccation cracking. The soil must be within the the proper range of water content prior to compaction.

Small adjustments in water content are easily made just prior to compaction. Large (more than about 3 percentage points in water content) adjustments in water content should not be made just prior to compaction because the soil must be given time to moisten or dry uniformly. To make large adjustments in water content, the author recommends wetting or drying the soil in a separate moisture conditioning area. It is important to: (1) mix the soil during wetting or drying (e.g., with tilling equipment), and (2) allow time for the soil to equilibrate. After moisture adjustment and equilibration, the soil can be stockpiled or hauled off for placement and compaction.

Surface Preparation. If the soil is placed on a natural soil subgrade or miscellaneous fill, the subgrade should be adequately compacted. If the soil is placed on a previously-compacted lift, the surface should be scarified to a nominal depth of 25 mm (1 inch) prior to placing soil for the next lift. The reason for this is to blend lifts together and eliminate seepage along lift interfaces, which is crucially important (Herrmann and Elsbury, 1987).

The soil must be placed in lifts that are not too thick; otherwise, the lower portion of the lift will not be well compacted and lifts will not be effectively bonded. Loose lifts no thicker than 230 mm (9 inches) and compacted lifts be no thicker than approximately 150 mm (6 inches) are recommended.

Compaction. The best type of compactor in most instances is a heavy, footed roller with feet that fully penetrate a loose lift of soil. The roller must be heavy to ensure that adequate compactive energy is delivered to the soil. Fully-penetrating feet knead the soil over the full depth of the lift and help to bond lifts. The author recommends for most applications that the roller have a static weight of at least 40,000 lbs (18,000 kg) for compacting flat areas and 30,000 lbs (14,000 kg) for compacting liners on slopes. (A lighter roller is sometimes needed on slopes to prevent tearing or shearing the liner.) However, a balance must be struck between weight of the roller and water content of the soil (the drier the soil, the heavier the roller).

It is important that the compactor passes over a given area an adequate number of times. Dry density alone is not always a good indicator of compactive energy, especially for soils compacted wet of optimum. Mitchell et al. (1965) presented data showing that the hydraulic conductivity of soils compacted wet of optimum can be decreased by as much as 100 times if more energy is used in compacting the soil, even though the increased compactive energy does not measurably increase the dry density of the compacted soil. The minimum number of passes varies between soils and compactors, but it is the author's experience that good construction with footed rollers almost always requires at least 5 passes, and often more. The author defines a pass as one pass of the compactor over an area, not one pass of an axle. Some compactors have footed drums on the front and rear and compact the soil twice with one pass, although feet should not impact the same spot with the front and rear drums. Other compactors have just one drum and compact the soil just once per pass. More passes are likely needed with a roller that has a single drum.

Most engineers rely upon in situ measurements of the dry density of the compacted soil to ensure that a lift of soil has been adequately compacted. It is usually convenient to measure the water content at the same time that the density is measured (because one must know the water content to calculate the dry density from a measurement of total, or wet, density). Daniel and Benson (1990) describe a methodology for establishing water content-density criteria in the laboratory. One first determines the water content-density requirements necessary to meet hydraulic conductivity goals (see Figure 1). However, one must consider all relevant factors, including (depending on the project) shear strength, ductility, potential for desiccation damage, and possibly others, depending upon site-specific objectives.

<u>Protection</u>. Soil must be protected from desiccation cracking and frost damage. Often, a smooth-drum roller is used to compact the surface of a completed lift. This forms a hard "skin" which helps to minimize desiccation.

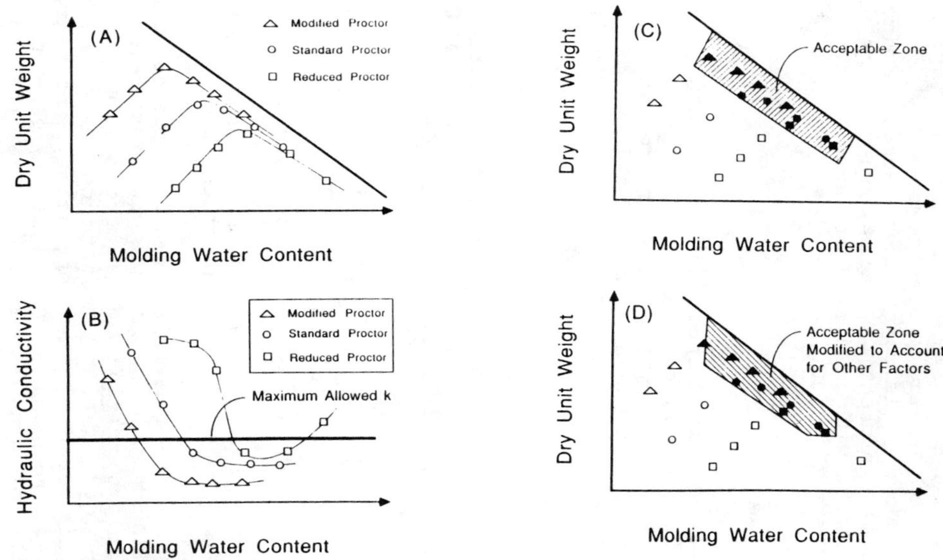

Figure 1. Procedure for Defining Acceptable Range of Water Content and Dry Unit Weight (from Daniel and Benson, 1990). (A) Compaction curves are developed with compactive energies spanning the range expected in the field; (B) Compacted soils are permeated and hydraulic conductivity plotted vs. molding water content; (C) Compaction curves are replotted using different symbols for soils with acceptable and unacceptable hydraulic conductivities, and an Acceptable Zone is defined; and (D) Acceptable Zone is modified for to account for other factors, e.g., local construction practices or requirements for shear strength.

The smooth skin also helps to shed water should a rainstorm occur. The smooth surface should be roughened with a disc prior to placing the next lift. In addition to smooth-rolling the surface, the soil may be periodically moistened and/or covered temporarily with a sheet of plastic. Good QA is essential to ensure minimal desiccation. The best protection against frost damage is to cover the soil liner with soil or (for landfill liners) solid waste to protect the liner.

Test Pad

To verify that the materials and methods of construction will produce the desired results, construction of a test pad (Figure 2) is recommended (U. S. EPA, 1985). The in situ hydraulic conductivity of the field-compacted soil should be measured (Daniel, 1989). Once a suitably low hydraulic conductivity has been verified, QA should be used to ensure that the actual soil liner is built to standards that equal or exceed those used to build the test pad. The test pad also affords an opportunity to evaluate the QC tests, to calibrate measurement devices, and to understand better the QA requirements. Test pads also provide a powerful means for proving that non-standard materials or construction procedures will work. If there is a substantial change in the soil material or construction methodology, a new test pad may be needed.

In cases where a test pad is not used, the author strongly encourages construction of a "demonstration strip." Such a strip has a width nominally equal to the width of the compactor, is 3 to 4 lifts thick, and is 2 to 3 times longer than the width. As the strip is constructed, QC tests (such as moisture-density tests) are performed to calibrate the equipment and to make sure there are no problems that would require altering the testing methodology. Also, construction of the demonstration strip enables the contractor, engineer, owner/operator, and regulatory officials to come to a common understanding of what is expected. After the strip is completed, a backhoe should be used to dig through the strip. The exposed soil should be inspected for evidence of voids, large pores between remnant clods, signs of poor bonding between lifts, and other visible problems.

Quality Control (QC) Testing

Materials. The recommended tests and frequency of testing are listed in Table 1. These tests were derived

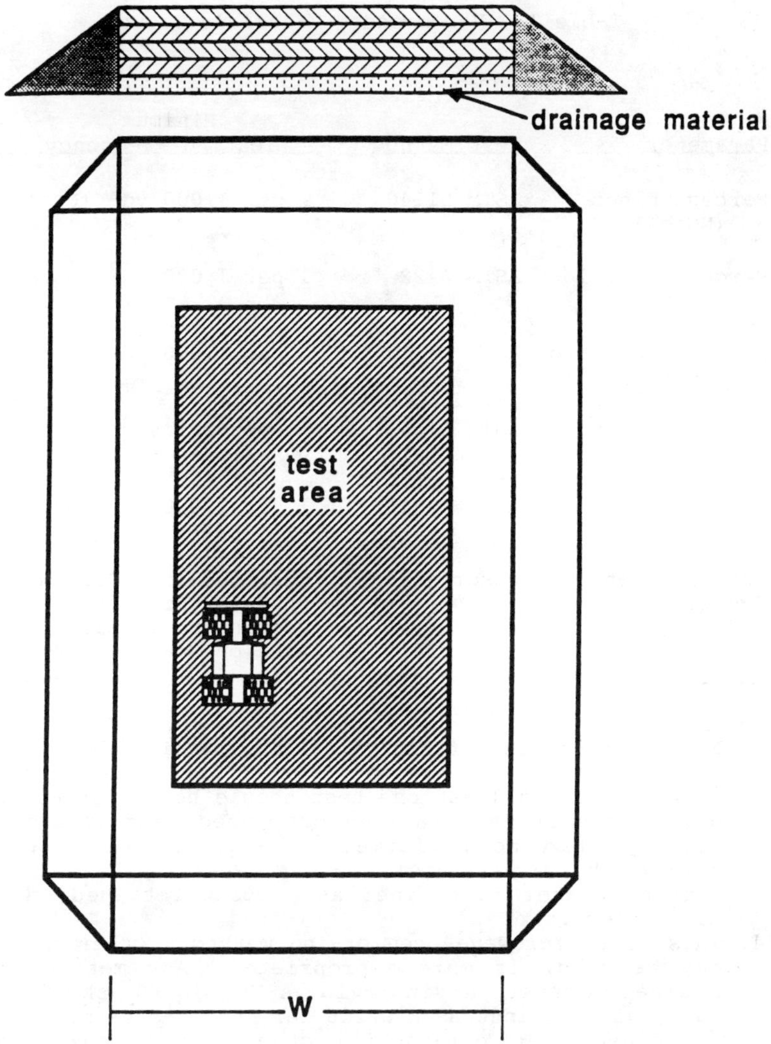

Figure 2. Test Pad for Verifying that Materials and Methods of Construction Will Produce Desired In Situ Hydraulic Conductivity.

Table 1. Recommended Materials Tests

Parameter	Test Method	Minimum Testing Frequency
Percent Fines (Note 1)	ASTM D1140	1 per 1,000 yd^3 (Note 2)
Percent Gravel (Note 3)	ASTM D422	1 per 1,000 yd^3 (Note 2)
Liquid & Plastic Limits	ASTM D4318	1 per 1,000 yd^3 (Note 2)
Water Content	ASTM D4643 (Note 4)	1 per 200 yd^3 (Note 2)
Water Content (Note 5)	ASTM D2216	1 per 1,000 yd^3 (Note 5)
Construction Oversight	Observation	Continuous in Borrow Pit on Major Projects; Continuous in Placement Area on Smaller Projects

Notes:
1. Percent fines is defined as percent passing the No. 200 sieve.
2. In addition, at least one test should be performed each day that soil is excavated or placed, and additional tests should be performed on any suspect material observed by QA personnel.
3. Percent gravel is defined as percent retained on the No. 4 sieve.
4. This is a microwave oven drying method. Other methods may be used, if more appropriate. Any method used besides direct drying via ASTM D2216 should be calibrated against ASTM D2216 for the on-site soils.
5. Microwave oven drying and other rapid measurement methods may involve systematic errors. Conventional oven drying (ASTM D2216) is recommended on every fifth sample taken for rapid measurement. The intent is to document any systematic error in rapid water content measurement.
6. 1000 yd^3 = 760 m^3.

primarily from recommendations published by Gordon et al. (1984), Goldman et al. (1988), and the New York Department of Environmental Conservation (1988). The tests can be performed on samples taken from the borrow area, pretreatment area, or final placement area, although the author recommends sampling in the borrow area because it is best to focus control efforts in the borrow area where adjustments in material selection procedures are most convenient.

Surface Preparation. The recommended tests and observations for surface placement are listed in Table 2. Preparation of a previously-compacted lift, e.g., via scarification of the surface, is best evaluated with observations rather than tests.

Placement of Soil. The loose lift thickness can be checked visually at the working face of placement. If the thickness is questionable, a hole can be dug in the soil and the loose lift thickness checked.

Many contractors control fill levels with grade stakes. Care must be taken to remove the grade stakes and to repair the holes left by the grade stakes. The QA inspector should make sure grade stakes are not buried in the liner. The holes left by grade stakes should be packed with soil liner material or bentonite that is tamped into the hole in layers with a rod.

Compaction of Soil. The tests and observations recommended for ensuring that the soil is properly compacted are listed in Table 3. The basis for the recommended frequency of testing is Gordon et al. (1984). Periodic calibration checks are recommended for tests performed with microwave ovens, nuclear devices, drive rings, or other apparatus that may introduce a small, systematic bias in the test results. Systematic measurement errors, particularly for water content, which is a crucial construction variable, must be documented.

Some people recommend laboratory hydraulic conductivity tests on "undisturbed" soil taken with thin-walled sampling tubes from each lift (Gordon et al., 1984, for example). Such tests have been omitted from Table 3 because: (1) the tests have produced grossly misleading results for poorly constructed liners (Daniel, 1984; Day and Daniel, 1985; and Elsbury et al., 1990); (2) the tests often take several days or more to complete, which minimizes their utility as a QC tool; and (3) for gravelly soils, good samples are almost impossible to obtain. While low k from undisturbed samples is necessary for a well-built liner, low k from tube samples is not

Table 2. Recommended Tests and Observations on Subgrade Preparation

Parameter	Test Method	Minimum Testing Frequency
Percent Compaction (Note 1)	ASTM D2922 or ASTM D1556 or ASTM D2937 or ASTM D2167	1 per acre (Note 2)
Compaction Curve	ASTM D698 (Note 3)	1 per 5 acres
Preparation of Previously Compacted Lift	Observation	Full Coverage

Notes:

1. Percent compaction is defined as the dry density of the compacted soil divided by the maximum dry density measured in the laboratory with a specified method of compaction. The test methods listed are for measurement of the dry density of the compacted soil.
2. In addition, at least one test should be performed each day the construction personnel prepare subgrade by compaction.
3. Other laboratory compaction methodologies are often employed.
4. 1 acre = 0.4 ha.

Table 3. Recommended Tests and Observations on Compacted Soil

Parameter	Test Method	Minimum Testing Frequency
Water Content (Note 1)	ASTM D3017 or ASTM D4643	5/acre/lift (Note 2)
Water Content (Note 3)	ASTM D2216	1/acre/lift (Note 3)
Density (Note 4)	ASTM D2922 or ASTM D2937	5/acre/lift (Note 2)
Density (Note 5)	D1556	1/acre/lift (Note 5)
Number of Passes	Observation	1/acre/lift (Note 2)
Construction Oversight	Observation	Continuous

Notes:

1. ASTM D3017 is a nuclear method and D4643 is microwave oven drying. Direct water content determination (ASTM D2216) is the standard against which nuclear, microwave, or other methods of measurements are calibrated for on-site soils.
2. In addition, at least one test should be performed each day soil is compacted and additional tests should be performed in areas for which QA personnel have reason to suspect inadequate compaction.
3. Every fifth sample tested with ASTM D3017 or D4643 should be also tested by direct oven drying (ASTM D2216) to aid in identifying any significant, systematic calibration errors with D3017 or D4643.
4. ASTM D2922 is a nuclear method and D2937 is a drive ring method. These methods, if used, should be calibrated against the sand cone (ASTM D1556). Alternatively, the sand cone method can be used directly.
5. Every fifth sample tested with D2922 or D2937 should also be tested (as close as possible to the same test location) with the sand cone (ASTM D1556) to aid in identifying any systematic calibration errors with D2922 or D2937. The sand cone method may be used in lieu of D2922 and D2937.
6. 1 acre = 0.4 ha.

sufficient to demonstrate low in-field hydraulic conductivity. While laboratory hydraulic conductivity tests on tube samples may provide interesting results that could potentially be useful, it is not clear to this author that such tests are worth the time and expense or that they provide data that are very meaningful.

Any holes made in the liner must be sealed. Quality assurance personnel should visually inspect the sealing of some of the holes made for QC testing.

Protection of a Completed Lift. This type of inspection is best done by visual observation to determine if adequate measures have been taken to protect each lift of soil from desiccation and freezing. Additional water content tests are recommended in cases where desiccation is suspected.

Sampling Pattern. Random sampling is recommended as the basic sampling pattern. The author suggests dividing the area to be tested into a grid and using a random number generator to determine sampling points. Random number generators are available in various microcomputer software packages. Hilf (1975, p. 303) describes a hand method of generating random numbers, and many mathematics handbooks contain tables of random numbers. The sampling pattern in each lift should be staggered from that in overlying and underlying lifts.

Erroneous Measurements. Occasionally, a mistake is made in a measurement. To determine if a mistake has been made, the test may be repeated several times immediately adjacent to the area where the questionable test was performed. There is no consensus on how many "passing" tests are needed to overrule one "failing" test, but most people seem to accept a need for 2 to 4 passing tests to invalidate a failing test.

Failed Tests. If for some reason a test fails, the area that does not meet specifications must be made to meet the specifications. A systematic pattern of testing should be employed to identify the extent of the area requiring further work.

Outliers. Soil is a variable material. It is naive to think that a soil liner will meet specifications at each and every point in the liner (if enough samples were taken, surely some out-of-specification material would be found). Indeed, it is for this very reason that soil liners are built in multiple lifts, i.e., so that a few imperfections will not matter. It is recommended that a certain percentage of test results be allowed to deviate from specifications, provided the failures are randomly distributed spatially and are not outrageous deviations. There is no consensus on what percentage of outliers is acceptable, but the author suggests that 2 percent outliers (and probably more) is small enough to be of no consequence.

Conclusions

Good construction quality assurance (QA) for compacted soil liners is a complex task with many requirements. Although some aspects can be quantified, many cannot. There is both art and science to good QA for soil liners. The author has attempted to describe a framework for good QA. However, he emphasizes that there is no consensus in the profession about precisely which tests should be performed and at what frequency. The quantitative suggestions made herein should be regarded as guidelines or starting points for thought and discussion, not as absolute values to apply indiscriminately on all projects.

Acknowledgments

The author has benefited from information and insight provided by many individuals, all of whom he sincerely thanks. Special acknowledgment is given to Gordon Boutwell and Stephen Trautwein, who have been particularly generous in sharing their wealth of knowledge on this subject.

References

Albrecht, K. A., and K. Cartwright (1989), "Infiltration and Hydraulic Conductivity of a Compacted Earthen Liner," Ground Water, Vol. 27, No. 1, pp. 14-19.

Boutwell, G. P., and Hedges, C. S. (1988), "Construction Quality Assurance for Compacted Clay Liners," Third Annual Ground Water Seminar, Louisiana Dept. of Environmental Quality, Baton Rouge, LA.

Daniel, D. E. (1984), "Predicting Hydraulic Conductivity of Clay Liners," Journal of Geotechnical Engineering, Vol. 110, No. 4, pp. 465-478.

Daniel, D. E. (1987), "Earthen Liners for Land Disposal Facilities," Geotechnical Practice for Waste Disposal '87, R. D. Woods (ed.), ASCE, New York, pp. 21-39.

Daniel, D. E. (1989), "In Situ Hydraulic Conductivity Tests for Compacted Clay," Journal of Geotechnical Engineering, Vol. 115, No. 9, pp. 1205-1226.

Daniel, D. E., and Benson, C. H. (1990), "Water Content-Density Criteria for Compacted Soil Liners," Journal of Geotechnical Engineering, Vol. 116, No. 12, in press.

Day, S. R., and D. E. Daniel (1985), "Hydraulic Conductivity of Two Prototype Clay Liners," Journal of Geotechnical Engineering, Vol. 111, No. 8, pp. 957-970.

Elsbury, B. R., Sraders, G. A., Anderson, D. C., Rehage, J. A., Sai, J. O., and D. E. Daniel (1990), "Field and Laboratory Testing of a Compacted Soil Liner," U. S. EPA, Cincinnati, Ohio, EPA/600/S2-88/067.

Goldman, L. J., Greenfield, L. I., Damle, A. S., Kingsbury, G. L., Northeim, C. M., and R. S. Truesdale (1988), "Design, Construction, and Evaluation of Clay Liners for Waste Management Facilities," U. S. Environmental Protection Agency, Washington, D. C., EPA/530/SW-86/007F.

Gordon, M. E., Huebner, P. M., and Kmet, P. (1984), "An Evaluation of the Performance of four Clay-Lined Landfills in Wisconsin," Proceedings, Seventh Annual Madison Waste Conference on Municipal and Industrial Waste, Madison, Wisconsin, pp. 399-460.

Gordon, M. E., Huebner, P. M., and T. J. Miazga (1989), "Hydraulic Conductivity of Three Landfill Clay Liners," Journal of Geotechnical Engineering, Vol. 115, No. 8, pp. 1148-1160.

Herrmann, J. G., and Elsbury, B. R. (1987), "Influential Factors in Soil Liner Construction for Waste Disposal Facilities," Geotechnical Practice for Waste Disposal '87, R. D. Woods (ed.), ASCE, New York, pp. 522-536.

Hilf, J. W. (1975), "Compacted Fill," in Foundation Engineering Handbook, H. F. Winterkorn and H.-Y. Fang (eds.), Van Nostrand Reinhold Co., New York, 751 p.

Johnson, G. W., Crumley, W. S., and G. P. Boutwell (1990), "Field Verification of Clay Liner Hydraulic Conductivity," Proceedings, Symposium on Performance, Construction, and Operation of Waste Disposal Facilities, ASCE, San Francisco, in press.

Lahti, L. R., King, K. S., Reades, D. W., and A. Bacopoulos (1987), "Quality Assurance Monitoring of a Large Clay Liner," Geotechnical Practice for Waste Disposal '87, R. D. Woods (ed.), ASCE, New York, pp. 640-654.

Mitchell, J. K., Hooper, D. R., and Campanella, R. G. (1965), "Permeability of Compacted Clay," Journal of the Soil Mechanics and Foundations Division, ASCE, Vol 91, No. SM4, pp. 41-65.

Oakley, R. E. (1987), "Performance of Earth-Lined Containment Systems," Geotechnical Practice for Waste Disposal '87, R. D. Woods (ed.), ASCE, New York, pp. 117-136.

New York State Department of Environmental Conservation (1988), "6 NYCRR Part 360, Solid Waste Management Facilities," Albany, NY.

U. S. Environmental Protection Agency (1985), "Draft, Minimum Technology Guidance on Double Liner Systems for Landfills and Surface Impoundments," EPA /530-SW-85-014.

U. S. Environmental Protection Agency (1986), "Technical Guidance Document: Construction Quality Assurance for Hazardous Waste Land Disposal Facilities," Office of Solid Waste, Washington, DC, EPA/530-SW-86-031.

GEOMEMBRANE CONSTRUCTION QUALITY ASSURANCE

Jean-Pierre Giroud[1], Member, ASCE, and Ian D. Peggs[2]

Abstract

This paper includes two parts. In the first part, an overview of construction quality assurance of geomembrane liners is presented. The second part is devoted to a discussion of improvements of the quality and performance of installed geomembranes which may result from measures taken to alleviate the influence of temperature on geomembrane behavior.

1. INTRODUCTION

1.1 Purpose of this Paper

The concept of geomembrane construction quality assurance was first developed in 1983. Since that date, the practice of geomembrane construction quality assurance has dramatically improved the quality of installed geomembrane liners. Today, geomembrane construction quality assurance is a well-established practice in the United States and it has been extended to other geosynthetics such as geonets, geotextiles, and geogrids. Nonetheless, there are still areas where progress can be made.

Progress in geomembrane construction quality assurance requires a good understanding of the behavior of geomembranes. Such understanding may result from data obtained in the testing of geomembranes, as well as from data collected in conducting forensic analyses of lining systems that have failed. This paper shows that data which are analyzed from the combined viewpoints of a polymer specialist and a geotechnical engineer can lead to

[1]Geotechnical Engineer and [2]Polymer Specialist, GeoSyntec Consultants (formerly GeoServices Inc. Consulting Engineers and GeoSyntec Inc.), Boynton Beach, Florida, USA.

practical guidance which may improve the practice of geomembrane quality assurance and, thus, the performance of geomembranes.

A number of papers and reports have been published on construction quality assurance of geomembranes and other geosynthetics (Giroud and Kastman, 1983; Fluet, 1985; Giroud and Fluet, 1986; USEPA, 1986; Fluet, 1987; USEPA, 1987; Wallace and Steinle, 1987; Matrecon, 1988; Koerner and Bove, 1989; Monteleone and DiPippo, 1989; Wallace, 1989; USEPA, 1989; Fluet, 1990), and it does not seem necessary to write another paper describing in detail this well-established practice. Instead, this paper is devoted to an overview of the practice of geomembrane construction quality assurance, followed by a detailed study of an area in which progress can be made.

1.2 Organization of this Paper

The first part of this paper is devoted to a general overview of geomembrane construction quality assurance, including:

- definitions and goals (Section 2.1);
- scope (Section 2.2);
- seam testing (Section 2.3);
- cost and benefits (Section 2.4);
- historical perspective (Section 2.5); and
- state of practice (Section 2.6).

It should be noted that the overview addresses only geomembrane construction quality assurance, for the sake of simplicity. However, similar methods are used and similar comments could be made for other geosynthetics, as well as for other materials used in the construction of lining systems for waste disposal facilities.

The second part of this paper discusses the potential for improvement in the quality and performance of installed geomembranes using, as an example, measures that can be taken to alleviate the influence of temperature on the behavior of geomembranes. The discussion addresses the following topics:

- overview of effects of temperature on geomembranes (Section 3.1);
- temperature of geomembranes in the field (Section 3.2);

- effect of temperature on geomembrane tensile characteristics (Section 3.3);
- irreversible geomembrane contraction (Section 3.4);
- geomembrane thermal expansion-contraction (Section 3.5);
- contraction of a geomembrane resting on soil (Section 3.6);
- wrinkling of a geomembrane resting on soil (Section 3.7);
- differential expansion of adjacent panels (Section 3.8); and
- practical recommendations (Section 3.9).

2. OVERVIEW OF CONSTRUCTION QUALITY ASSURANCE

2.1 Definition and Goals of Geomembrane Construction Quality Assurance

Geomembrane construction quality assurance is a service, provided to the owner of a facility by a team independent from the geomembrane installer, which consists of monitoring the construction activities and quality control measures taken by the installer to verify and document that the geomembrane is installed in conformity with the design and specifications. The goals of geomembrane construction quality assurance can be summarized as follows:

- verify and document through conformance testing that the geomembrane delivered to the site meets the specifications;
- verify and document through a variety of monitoring activities that the geomembrane is installed according to design and installation specifications; and
- verify and document that materials adjacent to the geomembrane are placed according to placement specifications that are aimed at minimizing the risk of damage to the geomembrane.

These three goals can be summarized by three words: *conformance*, *integrity*, and *survival* of the geomembrane. These three conditions are necessary (but not sufficient) to ensure the long-term performance of the geomembrane.

Integrity of the geomembrane is a subtle notion. In the early 1980's the goal of integrity consisted essentially of minimizing the number of holes in the geomembrane due to punctures and improper seaming. In 1990, this goal is virtually achieved in most geomembrane installations subjected to adequate construction quality assurance. However, forensic analyses of several geomembrane failures have shown that geomembranes with apparent integrity at the time of installation have eventually failed as a result of construction techniques that were not identified as inadequate during the implementation of construction quality assurance. For example: overheating or excessive grinding during seaming may weaken the geomembrane and affect its long-term performance; or, installing the geomembrane under high ambient temperature may cause large tensile stresses and cracking of the geomembrane when the temperature decreases and the geomembrane contracts.

Therefore, progress can be made in geomembrane installation if design and installation specifications address all aspects that govern the overall integrity and durability of the geomembrane, and not only those measures aimed at minimizing the number of holes.

2.2 Scope of Construction Quality Assurance

The construction quality assurance process usually starts with the preparation of a construction quality assurance plan. This document should be prepared by the design engineer, or by a specialist working in coordination with the design engineer. The construction quality assurance plan should be developed in conjunction with the project specifications because both documents are interrelated and their mutual consistency is necessary to obtain a high quality installation.

The construction quality assurance plan is typically a 50- to 100-page site-specific document which describes in detail all matters pertaining to the construction quality assurance process, as discussed by Giroud and Fluet (1986). In particular, the construction quality assurance plan describes the construction quality assurance activities, which, for geomembranes, include:

- reviewing the design, in particular the design drawings, to verify clarity and completeness, as well as to gain a good knowledge of the project;
- reviewing the specifications to verify clarity and completeness, and to gain a detailed knowledge of the materials and construction procedures to be used at the site;

- reviewing the construction quality assurance plan to gain a thorough understanding of the construction quality assurance procedures and to evaluate their adequacy and feasibility;
- submitting comments to the author of the construction quality assurance plan if it appears that some of the procedures should be changed;
- monitoring and documenting the unloading of trucks delivering geomembrane rolls or fabricated panels to the site;
- monitoring the handling and on-site storage of geomembrane rolls or fabricated panels;
- recording the manufacturing roll and batch numbers of geomembrane rolls or fabricated panels delivered to the site;
- selecting samples from geomembrane rolls or fabricated panels delivered to the site for conformance testing;
- affixing a code number to samples and recording the manufacturing numbers of the rolls from which samples are taken (i.e., manufacturing numbers of the specific rolls in the case of a fabricated panel);
- labeling, packaging, and shipping samples to an independent laboratory for conformance testing;
- interpreting laboratory test results in accordance with the specifications and accepting or rejecting delivered rolls or fabricated panels;
- reviewing results of topographical surveys to verify that lines and grades of the site as constructed are consistent with the design;
- obtaining a written acceptance of the subgrade by the geomembrane installer;
- evaluating and documenting weather conditions (temperature, wind) for geomembrane placement, and informing the on-site owner's representative if requirements for weather conditions are not met, so the on-site owner's representative can decide to stop geomembrane placement;

- monitoring and documenting geomembrane placement as well as conditions of panels as placed;
- recording locations of installed panels and checking that panels have been installed in accordance with the design;
- monitoring trial seaming done prior to each seaming sequence to evaluate the seaming crew and equipment;
- evaluating and documenting trial seam test results in accordance with the specifications and accepting or rejecting crews and/or equipment;
- evaluating and documenting the suitability of weather conditions (temperature, wind, humidity) for seaming, and informing the on-site owner's representative when weather conditions do not meet the specifications, so the owner's representative can decide to stop geomembrane seaming;
- monitoring the seaming of the geomembrane;
- affixing a code number to each seam and recording data such as name of seamer, machine used, date and time of seaming, and ambient temperature;
- monitoring and documenting nondestructive testing done to evaluate continuity of all seams;
- selecting locations where geomembrane samples will be taken to conduct destructive testing;
- monitoring the cutting out of samples by the geomembrane installer's crew;
- affixing a code number to each sample and recording sample locations and any pertinent observations made during sampling;
- monitoring the cutting of the sample in three parts, one for the geomembrane installer, one for archiving, and one for testing by the independent laboratory;
- monitoring and documenting the field seam destructive tests performed by the geomembrane installer;

- labeling, packaging, and shipping samples to the independent laboratory for destructive testing;
- interpreting laboratory test results and accepting or rejecting seams in accordance with the construction quality assurance plan;
- monitoring and documenting the patching of holes due to sampling;
- monitoring and documenting the nondestructive testing of the seams associated with patching;
- monitoring and documenting the repair of the rejected seams and the nondestructive testing of the repairs;
- monitoring and documenting destructive testing related to seam repair, as indicated above for regular seams;
- performing systematic visual observation of the entire surface of the geomembrane to locate and document defects and indicating for each defect the type of repair that is required;
- monitoring and recording the repair of defects and the nondestructive testing of the repairs;
- recording the location and the nature of all repairs; and
- monitoring the placement of materials on top of the geomembrane.

The operations listed above must be implemented by a construction quality assurance team assigned to the site for the duration of lining system construction. During this time, the quality assurance team documents all operations, test results, decisions, and observations, using special forms or logs. In addition, each quality assurance monitor records all special observations or comments in a daily report.

The last phase of geomembrane construction quality assurance consists of preparing a final report, which includes:

- a brief description of the project, including type of facility, name of site, location, altitude, name of owner, design engineer, geomembrane installer, earthwork contractor, etc.;

- a detailed description of the lining system, including surface area, cross section, definition of all materials, etc.;
- a reference to the construction quality assurance plan;
- a copy of geomembrane material specifications;
- a copy of, or reference to, geomembrane manufacturer's quality control documentation;
- a general record of activities, such as dates of performance of quality assurance operations, number and names of quality assurance monitors, number and names of geomembrane installer's personnel, etc.;
- a complete photographic record including photographs of the site at different phases of construction, photographs of construction details, and photographs of all quality assurance operations;
- a copy of all forms and logs filled out by quality assurance monitors and of all their daily reports;
- a copy of all field and laboratory test results;
- a discussion of special problems encountered and their solutions;
- a copy of the written acceptance of the subgrade by the geomembrane installer;
- a quality assurance record drawing indicating panel code numbers, seam code numbers, dates of seaming and repairs, locations and nature of all repairs, locations of samples taken for destructive testing and date they were taken;
- a construction record drawing showing those construction details that are different from details presented in design drawings; and
- a statement that construction has been done in substantial accordance with the design (including modifications, if any, approved by the design engineer).

Only geomembrane construction quality assurance has been discussed in this section. However, similar activities and documentation would be required for other geosynthetics.

2.3 Seam Testing

Seam testing is one of the most critical steps in geomembrane construction quality assurance and deserves a special discussion. Seam testing includes nondestructive testing done in the field, and destructive testing done in the field and in the laboratory. Nondestructive testing is intended to evaluate the continuity of seams and it is performed on the entire seam length (with the exception of a few locations that are not accessible to nondestructive testing equipment). Destructive testing is intended to evaluate the ability of the seams to withstand tensile stresses (through measurement of strength and elongation), and it can only be performed on a limited number of samples. Methods for nondestructive testing are discussed in detail elsewhere (Giroud and Fluet, 1986; Matrecon, 1988; Koerner, 1989) and only destructive testing is discussed below.

It is important to evaluate seam strength and elongation for two reasons. First, evaluating seam strength and elongation is the most objective way to measure the quality of the seams; therefore, it is the most objective way to evaluate the work done by the seaming crew and to assess the integrity of the geomembrane liner. Second, it is important to verify that seams are as strong and ductile as possible to minimize the risk of seam failure when the overburden stresses are applied. Ideally, seams should be as strong and ductile as the geomembrane itself, but this is difficult to achieve if only because the presence of the seam causes a disturbance: a seam may overstress and/or weaken the geomembrane in its vicinity. Therefore, special criteria are needed. Two types of tests are carried out to measure seam strength: shear tests and peel tests. These tests have been thoroughly discussed in the literature (Matrecon, 1988; Koerner, 1989) and are not described here. Commonly specified seam strength criteria based on these two tests are given in Table 1, which is adapted from the National Sanitation Foundation Standard 54 (NSF, 1985). However, these criteria are not sufficient as discussed below. Criteria related to shear tests are discussed first, then criteria related to peel tests.

Failure in a shear test occurs after a certain apparent strain. (Note: the phrase "apparent strain" is used because it is not possible to measure an actual strain in the shear test due to the presence of the seam.) Failure of an HDPE geomembrane in a tensile test may occur at a low strain if the geomembrane has surface irregularities (such as gouges, creases, and notches) (Giroud, 1984d) or if the geomembrane is brittle for some reason. Therefore, if in a

Table 1. Seam Strength Criteria for HDPE and PVC Geomembranes. (Adapted from NSF (1985).)

Test	Geomembrane Type	Minimum Required Seam Strength	Alternative Requirement
Shear	HDPE < 2 mm	90% yield strength	none
	HDPE > 2 mm	75% yield strength	none
	PVC	80% break strength	none
Peel	HDPE	none	or FTB
	PVC	1.75 kN/m	or FTB

Note: Ten specimens are cut from a geomembrane sample. Five are tested in shear, and five in peel. There is no indication as to whether the NSF criteria apply to individual specimens or to average values for each set of five specimens. Typically, unless otherwise specified, the sample is deemed acceptable if at least four specimens pass the shear test and four pass the peel test (USEPA, 1989).

shear test an HDPE geomembrane fails in the vicinity of the seam with a low apparent strain (e.g., 50% compared to 700% or more for the parent material), it may be concluded that the geomembrane has been weakened by the seaming process (Peggs, 1990b). The following mechanisms may have caused the weakening:

- deep grinding gouges parallel to the direction of the seam (for those seams that require grinding of the geomembrane surface prior to seaming);
- deformations of the geomembrane (such as creases and notches) caused by overheating during seaming; and/or
- embrittlement of the geomembrane due to microstructural changes caused by thermal energy input during seaming.

In service, geomembranes that have been thus weakened may fail in tension or as a result of stress cracking. Indeed, geomembrane failure as a result of tension and/or stress cracking has often been observed at the edges of seams (Peggs and Carlson, 1989).

Failure in a peel test occurs after partial or total separation of the initially bonded area. A very wide range of values (from a few percent to 100%) have been observed for the ratio between the separated area and the initially bonded area (separation ratio). As seen by its absence from Table 1, there is often no requirement regarding the separation ratio. However, research has demonstrated that crazes, the precursors of stress cracks, can be initiated in the separated surfaces, possibly as a result of the large shear stresses generated during peeling (Peggs, 1987). Measurements on geomembrane specimens cut out from the area separated during a peel test have shown stress crack resistances approximately seven times less than those of intact geomembranes (Peggs, 1990b). (Stress crack resistance was measured by subjecting the specimens to constant loads less than the yield strength and comparing the times to failure, according to test method GRI-GM5 from the Geosynthetics Research Institute.) In the field, peeling can occur in a variety of situations such as at wrinkles in an expanding and contracting liner, or even by seam rotation during the application of shear stresses across the seam (Giroud, 1984d). In some cases, a seam is strong enough not to fail when subjected to peeling forces in the field, but a large fraction of the initially bonded area (e.g., 50%) separates prior to the development of the required peel strength. In such cases, the risk of craze development is high. Then stress cracking may occur, and the service life of the geomembrane may be reduced.

Another important consideration in seam testing is the mode of failure. In both the shear test and the peel test, failure should occur through the geomembrane, not through the seam. Such a failure characteristic is termed "film-tearing bond".

Based on the above discussion, the parameters that design engineers should consider using in HDPE geomembrane seam acceptance criteria are:

- shear strength expressed as a percentage of the geomembrane yield strength;
- shear test "apparent strain";
- location of shear failure (film-tearing bond or otherwise);
- peel strength expressed as a percentage of the geomembrane yield strength or shear strength;
- peel separation ratio; and
- location of peel failure (film-tearing bond or otherwise).

The actual values used for the parameters should be defined for each project by the design engineer depending upon the materials used and the performance criteria. For example, the following acceptance criteria have been used for the seams of a specific HDPE geomembrane on a specific project:

- shear strength greater than 90% of the geomembrane yield strength expressed in units of force per unit width;
- shear test apparent strain greater than 50%;
- shear failure of the film-tearing bond type;
- peel strength greater than 60% of the geomembrane yield strength expressed in units of force per unit width;
- peel separation ratio less than 10%; and
- peel failure of the film-tearing bond type.

Many combinations of the parameters listed above may be used and the authors encourage geomembrane installers, design engineers, and testing experts to meet and discuss their experiences with different criteria in order that high quality standards may be developed for HDPE geomembranes, as suggested above, and for other geomembranes as well.

2.4 Cost and Benefits of Construction Quality Assurance

Based on data from actual landfills, the cost of geomembrane quality assurance varies from 5 to 20% of the cost of the geomembrane (material plus installation). The wide range results from the variety of geomembranes and lining systems, as well as the various degrees of construction quality assurance that are implemented. It appears that the cost of geomembrane construction quality assurance is not negligible. However, benefits are very substantial, as discussed below.

During the period 1983-1990, the quality of installed geomembranes and other geosynthetics has dramatically increased in the United States as a result of construction quality assurance. As discussed by Giroud and Fluet (1986), geomembrane construction quality assurance has been beneficial not only to owners of geomembrane-lined facilities, but to all parties involved, and to society at large through an increased protection of the environment against pollution. More specifically, Giroud and Bonaparte (1989) determined, on the basis of an analysis of actual installations by Giroud and Fluet (1986), that extensive construction quality assurance decreased the number of geomembrane defects by a factor of 30 compared to a geomembrane liner installed by a competent installer, but without quality assurance. With such a decrease, it is clear that geomembrane quality assurance is cost effective, and, more importantly, results in a large benefit in terms of protection of the environment. (It may be argued that, today, i.e., approximately five years after the study by Giroud and Fluet, the factor mentioned above would be less than 30 due to better seaming techniques and better quality control by geomembrane installers. However, it may also be argued that construction quality assurance has progressed in the past few years so that defects not found in 1986 would be found today.)

2.5 Historical Perspective of Geomembrane Quality Assurance

In the early 1980's, the quality of many geomembrane installations was less than it could and should have been. In the opening address of the International Conference on Geomembranes in 1984 it was stated that "...geomembranes are not always as successful as they could be. The main reason is that many people, when they use geomembranes, tend to neglect two important aspects, design and quality assurance, because they believe that geomembranes are absolutely impermeable in all circumstances. The fact that there is no such thing as an absolutely impermeable geomembrane cannot be emphasized enough... Of course, impermeability is a legitimate goal, especially for conservation of resources and protection of the environment which are two of the most important challenges of our times, but this goal cannot be achieved by just putting a geomembrane on the ground — proper design and quality assurance are required." (Giroud, 1984b).

Fortunately, the profession reacted aggressively and considerable progress has been made. A first attempt at a systematic guidance document on geomembrane construction quality assurance was the preparation of "Guide Specifications for Construction of Flexible Membrane Liners for Hazardous Waste Disposal Facilities" (Giroud and Kastman, 1983) as a part of a state-of-the-art review prepared for the United States Environmental Protection Agency (USEPA). In 1983-84, this document inspired the quality assurance of several specific installations where significant experience was gained.

Considerable relevant experience was also gained from failures. A number of partial or total failures of geomembrane-lined containment facilities were observed and documented for more than a decade and the various mechanisms of failure were organized by categories (Giroud, 1984a).

In the United States, a major impetus to quality assurance of geomembrane-lined hazardous waste disposal facilities was given by the publication by the USEPA of "Minimum Technology Guidance on Double Liner Systems for Landfills and Surface Impoundments — Design, Construction, and Operation" (USEPA,1985). One of the most notable aspects of this document is the requirement for extensive geomembrane construction quality assurance. Detailed requirements will be codified when the USEPA's proposed Liner and Leak Detection Rule is implemented (see USEPA, 1987).

By 1984, some large companies owning waste disposal facilities had retained engineering consultants to prepare site-specific construction quality assurance plans. In 1985-86, they started preparing construction quality assurance manuals, which are comprehensive generic documents to be used as guides for the preparation of site-specific construction quality assurance plans. This effort contributed significantly to the rationalization of geomembrane construction quality assurance and inspired the preparation by the USEPA of a technical guidance document entitled "Construction Quality Assurance for Hazardous Waste Land Disposal Facilities" (USEPA, 1986).

2.6 Geomembrane Construction Quality Assurance State of Practice

Today, in 1990, geomembrane construction quality assurance is a well-established practice in the United States. It is mandatory for geomembrane liners and caps used in hazardous waste disposal facilities and, in many states, it is also mandatory for geomembranes used for lining and capping municipal waste landfills. An increasing number of engineering consultants in the United States provide quality assurance services and an increasing number of laboratories provide the testing services related to geomembrane construction quality assurance. Although construction quality assurance began with geomembranes, it has been extended to all components of lining systems, which now often include a variety of geosynthetics (e.g, geonets, geotextiles, and geogrids, in addition to geomembranes) as well as natural materials (e.g., clay, sand, and gravel), composite materials such as fabric-contained bentonite panels, and also manufactured products such as pipes, sumps, etc.

Although the state of practice can be considered satisfactory, progress in geomembrane construction quality assurance may result from a better understanding of geomembrane behavior as discussed in the remainder of this paper, using as an example the influence of temperature on geomembranes.

3. INFLUENCE OF TEMPERATURE ON GEOMEMBRANES

3.1 Overview of Effects of Temperature on Geomembranes

The effects of temperature on geomembranes can be either reversible or irreversible. Two reversible effects are the change in geomembrane modulus with temperature and geomembrane thermal expansion-contraction. An irreversible effect is the geomembrane contraction that results from a variety of mechanisms mostly significant at elevated temperatures and sometimes referred to as dimensional instability. (There is no known threshold above which irreversible changes occur and below which they do not occur; however, changes occur faster, and possibly to a greater extent, as temperature increases.) Both reversible and irreversible effects may have a number of detrimental consequences on the performance of geomembranes as discussed below. Consequences related to contraction will be discussed first, then consequences related to expansion will be discussed.

Contraction of a geomembrane may be reversible (if it is due to a decrease in the ambient temperature) or irreversible (if it is due to dimensional instability). Contraction of a geomembrane liner can have noticeable effects such as the bridging of depressed areas or toes of slopes. If a geomembrane that is bridging an area is loaded (e.g., by filling a liquid impoundment), the seams will no longer be subjected predominantly to shear, as they would be on a flat area, but to a combination of shear and peel (Giroud, 1984d), which is more severe and may lead to seam failure. Another problem is that geomembrane liners are not totally free to contract due to friction with the supporting soil and/or because they are laterally anchored, or seamed to other panels; as a result, they develop tensile stresses due to constrained thermal contraction. When the temperature is very low, the geomembrane modulus is very high and the geomembrane becomes relatively brittle; the high tensile stresses resulting from geomembrane contraction added to service stresses, and combined with geomembrane microstructural defects, may then cause stress cracks to be initiated. Also, polymeric compounds have a low coefficient of thermal conductivity. The resulting difference in temperature between the geomembrane side exposed to the cold weather and the side in contact with the supporting soil causes differential contraction. This differential contraction tends to open any notch defects present at the geomembrane surface and causes them to propagate more deeply into the geomembrane, thereby causing stress cracking. (A semi-crystalline polymeric compound such as high density polyethylene (HDPE) is susceptible to stress cracking due to the existence of interfaces of relatively weak bonding between the crystalline areas (spherulites) and between the oriented lamellae within the spherulites. Polymeric compounds that are predominantly amorphous such as very low density polyethylene (VLDPE) and polyvinyl chloride (PVC) are intrinsically less susceptible to stress cracking. However, PVC geomembranes become susceptible to stress cracking when chemically degraded and plasticizers are removed.)

Expansion of a geomembrane liner at high temperatures typically causes extensive wrinkling of the geomembrane. Wrinkles are more apparent with HDPE geomembranes than with PVC geomembranes not because HDPE expands more, but because it is stiffer (See Sections 3.5 and 3.7). During construction, wrinkles impede seaming and sometimes result in openings ("fishmouths") that have to be cut and reseamed. Geomembrane wrinkles also impede the placement of drainage materials (such as gravel or geonet) on top of geomembranes to construct leachate collection layers and leakage detection layers. Wrinkles can also be snagged during construction operations, especially during the placement of materials overlying the geomembrane, thereby resulting in geomembrane damage. During operation of geomembrane-lined facilities, geomembrane wrinkles that are located under a drainage layer obstruct the flow of liquid in the drainage layer, whereas geomembrane wrinkles that are flattened or folded over by placement of overburden may be overstressed or stressed at a level that could initiate stress cracking.

Further discussions of the mechanisms mentioned above and recommendations to alleviate the resulting problems will be presented in the following sections.

3.2 Temperature of Geomembranes in the Field

In the field, exposed geomembranes may be subjected to a wide range of temperatures: from the lowest possible ambient temperatures in the winter (e.g., -50°C) to 90°C for black geomembranes exposed to the sun in the summer. In some areas of the western United States, a geomembrane temperature range of -35°C to 90°C can be observed at a given site. In the northeastern United States, geomembrane temperatures from -20°C in the winter to 75°C on a sunny summer day are common.

A layer of soil placed on the geomembrane drastically decreases the temperature range to which the geomembrane is exposed. A given variation of the soil surface temperature causes a variation of geomembrane temperature that decreases with increasing thicknesses of soil covering the geomembrane and increasing periods of the variation. For example, Table 2 gives the thickness of cover soil necessary to attenuate a variation of soil surface temperature of 30°C to a geomembrane temperature variation of 1°C.

Table 2. Attenuation Depth for a Temperature Variation of 30°C. (From Segrestin and Jailloux, 1988.)

Period of Variation	Attenuation Depth (m)
1 hour	0.15
24 hours	0.50
1 week	1.30
1 month	2.80
1 year	10.50

Note: The attenuation depth was calculated as the depth where the temperature variation is 1°C.

3.3 Effect of Temperature on Tensile Characteristics

It is well known that, when temperature increases, thermoplastic materials such as HDPE and PVC soften and, when temperature decreases, they become stiffer and even brittle. This phenomenon can be evaluated in the laboratory by measuring the influence of temperature on geomembrane tensile characteristics.

The influence of temperature can be summarized as follows:

- as temperature increases, the strength at yield (for HDPE) and the strength at break decrease to reach a value of zero at the melting point of the considered thermoplastic (approximately 105-120°C for PVC and 100-135°C for polyethylene, depending on density, 135°C corresponding to a density of 940 kg/m^3);
- as temperature increases, the strains at yield (for HDPE) and at break increase; and
- as temperature increases, the initial modulus and the secant modulus at a given strain (e.g., at ϵ = 100%) decrease even more markedly than the strength, because modulus equals strength (which decreases) divided by strain (which increases).

Data regarding the changes in strength, strain, and modulus of PVC and HDPE geomembranes with temperature are given in Figures 1, 2, and 3. It may be noted that the

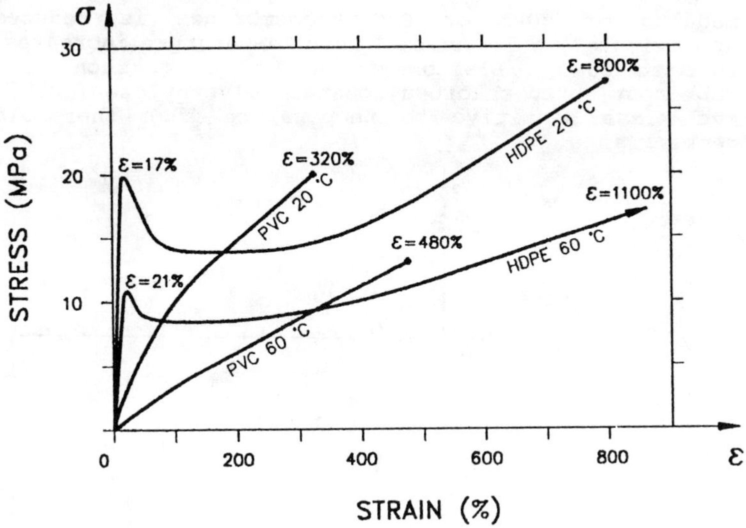

Figure 1. *Effect of Temperature on Tensile Characteristics of PVC and HDPE Geomembranes.* (From results of tests performed according to ASTM D638 with dumbbell type IV specimens by Matrecon (1988).)

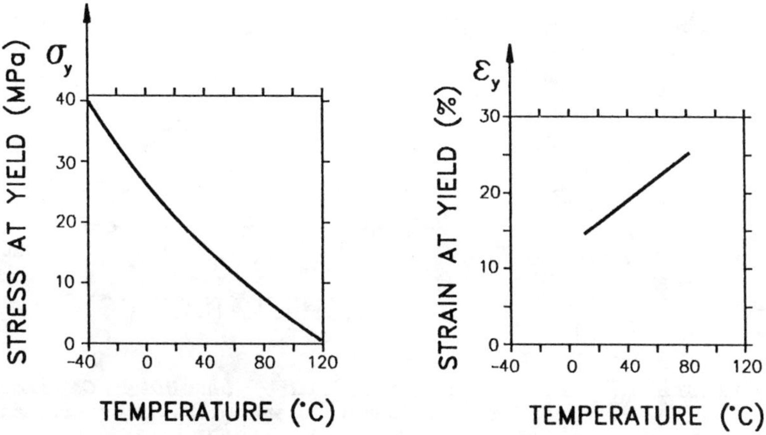

Figure 2. *Effect of Temperature on Yield Characteristics of HDPE Geomembranes.* These are typical values obtained with dumbbell specimens. (Combination of results published by Matrecon (1988), provided by manufacturers, and from other various sources.)

modulus of HDPE or PVC geomembranes is reduced by approximately a factor of 3 when temperature increases from 20°C to 60°C. Elastomeric geomembranes, such as butyl rubber or cured chlorosulfonated polyethylene (CSPE), are much less sensitive to temperature than thermoplastic geomembranes.

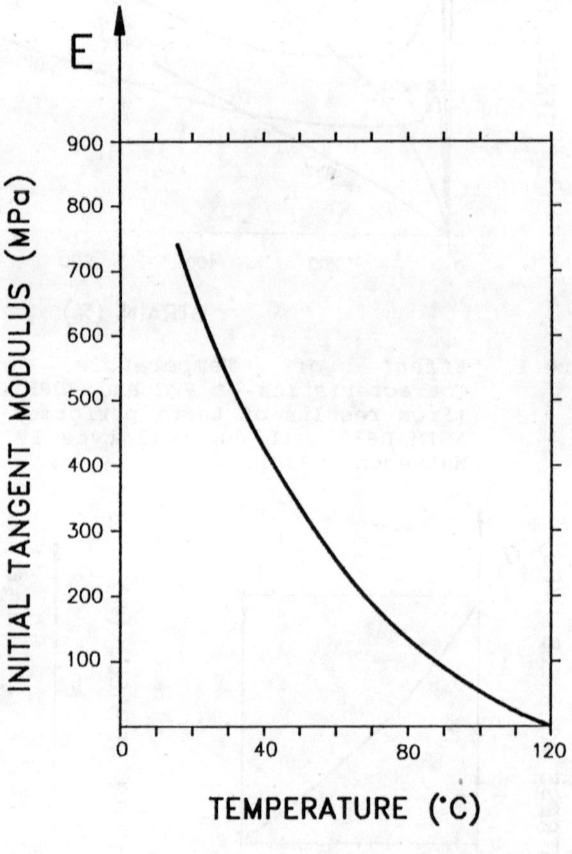

Figure 3. *Effect of Temperature on HDPE Geomembrane Modulus.* This curve was derived from results obtained by Matrecon (1988) using 13-mm wide strip specimens with an initial grip separation of 50 mm and a strain rate of 5 mm/min. The HDPE geomembranes used in the tests had specific gravities ranging between 0.94 and 0.96. According to Matrecon, higher values would have been obtained with the 25-mm wide strip specimen specified in ASTM D882.

3.4 Irreversible Shrinkage

Exposure to high temperatures causes loss of volatile components in geomembranes and relief of internal residual stresses. Both mechanisms result in irreversible contraction of the geomembrane, also called shrinkage. They are discussed below.

When a geomembrane is heated, its volatile components evaporate. This effect is insignificant in polyethylene geomembranes because they usually contain a negligible amount of volatile components (i.e., less than 1%). In contrast, PVC geomembranes typically contain approximately 35% by weight of plasticizers, which are rather volatile. Under extreme heat exposure, a PVC geomembrane may lose as much as one half of its plasticizers (Giroud, 1984c), or even more if the plasticizers are particularly volatile. A PVC geomembrane becomes significantly less flexible if it loses 25% of its plasticizers (i.e., if the plasticizer content decreases from the original 35% to 29%) and becomes extremely brittle if it loses half its plasticizers (i.e., if the plasticizer content decreases from the original 35% to 21%) (Giroud, 1984c). The volume change due to plasticizer loss can be evaluated using the following equation:

$$\Delta V/V = \lambda \, P_g \, \rho_g/\rho_p \qquad \text{(Equation 1)}$$

where: λ = proportion of plasticizer that has left the geomembrane (expressed as a fraction of the initial amount of plasticizer); P_g = proportion by weight of plasticizer initially present in the geomembrane (expressed as a fraction of the total weight of the geomembrane including the plasticizer); ρ_g = density of the geomembrane; and ρ_p = density of the plasticizer.

Shrinkage is typically expressed as a dimension change, $\Delta L/L$, which is related to the volume change as follows:

$$1 + \Delta L/L = (1+\Delta V/V)^{1/3} \qquad \text{(Equation 2)}$$

Combining Equations 1 and 2 leads to the following approximate relationship:

$$\Delta L/L = (1/3)\Delta V/V = \lambda \, P_g \, \rho_g/(3\rho_p) \qquad \text{(Equation 3)}$$

Using typical values such as $P_g = 0.35$, $\rho_g = 1{,}250$ kg/m^3, and $\rho_p = 1{,}050$ kg/m^3, the following value is obtained for the linear shrinkage:

$$\Delta L/L = 0.13 \, \lambda \qquad \text{(Equation 4)}$$

For example, if a PVC geomembrane loses 10% of its

plasticizers (which is not too detrimental to its flexibility), the resulting contraction is 0.013, i.e., 1.3% in all directions, which is not negligible. A similar calculation shows that an HDPE geomembrane that loses 10% of its volatile components shrinks by less than 0.03%, which is negligible.

During the manufacturing of a geomembrane, several operations (such as extrusion, calendering, and reeling up the geomembrane roll) tend to increase the length of the geomembrane and to orientate the polymer molecules. In other words, internal residual tensile stresses are stored in the geomembrane. When the geomembrane is heated (which increases the mobility of the molecules) these stresses are relieved. As a result the geomembrane contracts and this contraction is irreversible since the level of internal stresses decreases during the processes. There is no known temperature threshold beyond which this phenomenon occurs and below which it does not occur. However, it is known that sun-generated heat is sufficient to cause significant irreversible shrinkage to most geomembranes. The authors do not have experimental data on the magnitude of this phenomenon. However, the first author has been involved in a project where a PVC manufacturer indicated that the amount of tension stored in a PVC geomembrane as a result of the manufacturing process corresponded to a constrained contraction on the order of 2% to 4% in the roll direction.

In the field, contraction cannot take place if the geomembrane is not free to move due to geomembrane-soil friction or because the geomembrane is anchored at its periphery. If this is the case, the geomembrane develops tensile stresses corresponding to the constrained contraction.

Manufacturers' and NSF Standard 54 (NSF, 1985) require that the dimensional instability of an HDPE geomembrane when exposed to a temperature of 100°C for 15 minutes (ASTM D1204) not exceed 3%. Such a requirement is relatively easy to meet but it does not imply that there will be no further change in dimensions if the material is heated for a longer period. Figure 4 shows that HDPE geomembranes continue to contract even after 10,000 hr exposure to 100°C. Figure 4 also shows that the dimensional changes in the two principal directions of a given geomembrane may be very different. With some geomembranes, in the first few hours of testing, there may even be expansion in one direction and contraction in the other.

PVC geomembranes are less dimensionally stable than HDPE geomembranes. NSF Standard 54 (NSF, 1985) requires that the dimensional instability of a PVC geomembrane when exposed to a temperature of 100°C for 15 minutes (ASTM 1204) not exceed 5%. Comparing this requirement to the 3%

requirement for HDPE geomembranes mentioned above may be misleading: PVC geomembranes are probably 3 to 5 times more dimensionally unstable than HDPE geomembranes.

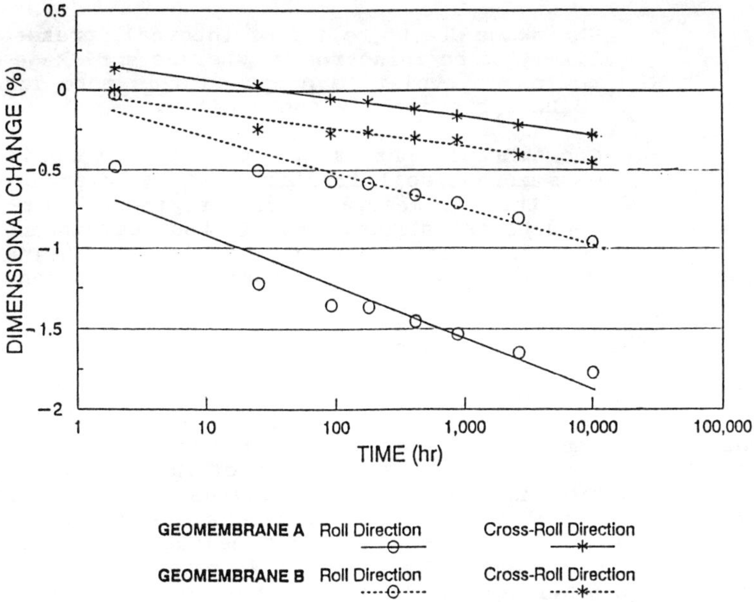

Figure 4. *Dimensional Changes in Two HDPE Geomembranes (A and B) at a Constant Temperature (100°C)*. The tests were conducted according to ASTM D1204 with square specimens, 250 mm x 250 mm. For both geomembranes, the machine direction is the direction of extrusion as well as the roll length direction.

The discussion presented above can be summarized as follows:

- Irreversible geomembrane shrinkage is due to loss of volatile components and relief of internal stresses.

- Shrinkage due to loss of volatile components is insignificant in the case of HDPE geomembrane but may be between 0.5% and 2% for PVC geomembranes (according to Equation 4).

- Shrinkage due to relief of internal stresses is difficult to evaluate. It can be assumed to be between 2% and 4% for exposed PVC geomembranes (in the roll direction) and is probably less than 1% for exposed HDPE geomembranes. Smaller values are expected for geomembranes covered by a layer of soil.

- Shrinkage due to relief of internal stresses is likely to be anisotropic, whereas shrinkage due to loss of volatile components is more likely to be approximately isotropic.

- Geomembrane shrinkage may be impeded by geomembrane/soil friction and/or by anchorage of the geomembrane in a trench or to appurtenant structures. If the geomembrane is not free to contract, the above shrinkage values result in corresponding tensile stresses.

3.5 Thermal Expansion-Contraction

Geomembranes exhibit significant reversible changes in dimensions when the temperature varies: they expand when the temperature increases and contract when the temperature decreases. The mechanism of thermal expansion and contraction is reversible as long as the geomembrane can move freely and is not irreversibly affected by temperature (i.e., when the dimensional instability component discussed in Section 3.4 remains relatively small). In other words, if, after expansion or contraction caused by a temperature change, the temperature comes back to its initial value, the geomembrane specimen recovers its original length.

The coefficient of thermal expansion-contraction of a geomembrane is defined as follows:

$$\alpha_{CTE} = (\Delta L/L) \;/\; \Delta T \qquad \text{(Equation 5)}$$

where: ΔL = change in length of a geomembrane specimen; L = initial length of the specimen; and ΔT = temperature difference that causes the change in length.

The terminology "coefficient of thermal expansion" (CTE) is generally used instead of coefficient of thermal expansion-contraction. This is acceptable, because there should be no quantitative difference between expansion and contraction since the mechanism is reversible. Strictly speaking, it is preferable to say "coefficient of linear expansion" in order to prevent confusion with the "coefficient of volume expansion", which quantifies the change in volume. However, with geomembranes, only the linear coefficient is used and, therefore, the adjective linear is usually omitted.

The coefficient of thermal expansion of a geomembrane is obtained by measuring the change in length of a geomembrane specimen under various controlled temperatures. According to ASTM D696, the specimen should have a length between 50 and 125 mm, and should be narrow enough to be contained in a tube with an inside diameter of 15 mm; when it expands, the specimen pushes a rod that activates a dial gauge. Another method, often used in materials testing laboratories but for which no testing standard exists to the authors' knowledge, consists of using a thermal mechanical analyzer, which requires very small cubic specimens, typically 1 mm^3. Extensive testing on HDPE geomembranes by the second author shows that:

- Most geomembranes are not rigid enough to be tested using the ASTM D696 procedure. The specimens tend to bend inside the tube instead of pushing the rod. Consequently, inaccurate results may be obtained.

- Test results obtained with the small cubic specimens are scattered and not consistently reproducible. This may be due to the fact that these small specimens are overly affected by microstructural variations within the geomembrane.

In contrast, consistent test results are obtained with large square specimens, e.g., 200 x 200 mm on which measurements are made in both principal directions, the roll direction and the cross-roll direction, and this methodology is recommended by the authors. Measurements can be made between bench marks on these specimens according to test methods that describe how to make precise length measurements, such as ASTM D696, D1042, and D1204. The only difficulty in following the methodology recommended by the authors is that the geomembrane specimens tend to curl when they are subjected to heat; they should be allowed to curl freely during exposure to temperature, but they must be kept flat for length measurement.

A wide range of values of coefficients of thermal expansion are mentioned in the literature: 7×10^{-5} to 7×10^{-4}/°C for both PVC and HDPE (Koerner, 1989; Matrecon, 1988; Peggs et al., 1990a). It should be noted that there is no significant difference between the coefficients of thermal expansion of HDPE and PVC geomembranes. A rather wide range of values is given because the coefficient of thermal expansion depends on several parameters:

- polymeric compound and cell structure;

- manufacturing process;

- direction (e.g., roll direction and cross-roll direction); and
- temperature range (e.g., the coefficient of thermal expansion is usually different between, for example, 0° and 20°, or 20° and 40°, or 0° and 40°).

In some situations, it may be necessary to take into account the fact that the coefficient of thermal expansion varies with temperature. In such cases, a graphic representation of thermal expansion is useful. Figure 5 shows the result of a test conducted on an HDPE geomembrane. Figure 5a shows the relative change in dimension of the specimen using the length at -40°C as a basis, and Figure 5b shows the coefficient of thermal expansion derived from the curves of Figure 5a. From a mathematical standpoint, the curves of Figure 5b are derivatives of the curves of Figure 5a. Thus, between -40°C and 0°C, the coefficient of thermal expansion is constant because the curves of relative change in dimension are linear.

It should be noted that the use of the specimen length at -40°C as a basis for the curves in Figure 5a is arbitrary. However, using any other length would give basically the same results since the specimen lengths at various temperatures differ by less than 1.5%.

Polymeric compounds have coefficients of thermal expansion larger than those of other construction materials such as concrete, steel, and glass. Typical values of the coefficients of thermal expansion of these materials at 20°C are:

- 1.0 to 1.1 x 10^{-5}/°C for steel;
- 1.0 to 1.4 x 10^{-5}/°C for concrete; and
- 0.3 to 1.0 x 10^{-5}/°C for glass.

The large value of the coefficient of thermal expansion of geomembranes can significantly affect their performance in the field. Precautions can be taken during geomembrane installation to alleviate potential performance problems.

In most cases, geomembranes are installed during the day (i.e., at a temperature higher than during the night), and they are not installed in cold weather. As a result, geomembranes are typically installed at a relatively high temperature and they will tend to contract when exposed to relatively low temperatures during the night and/or in the winter. The tendency of the geomembrane to contract may generate excessive tensile stresses if the geomembrane has not been installed with sufficient slack. Therefore, all geomembranes should be installed with slack.

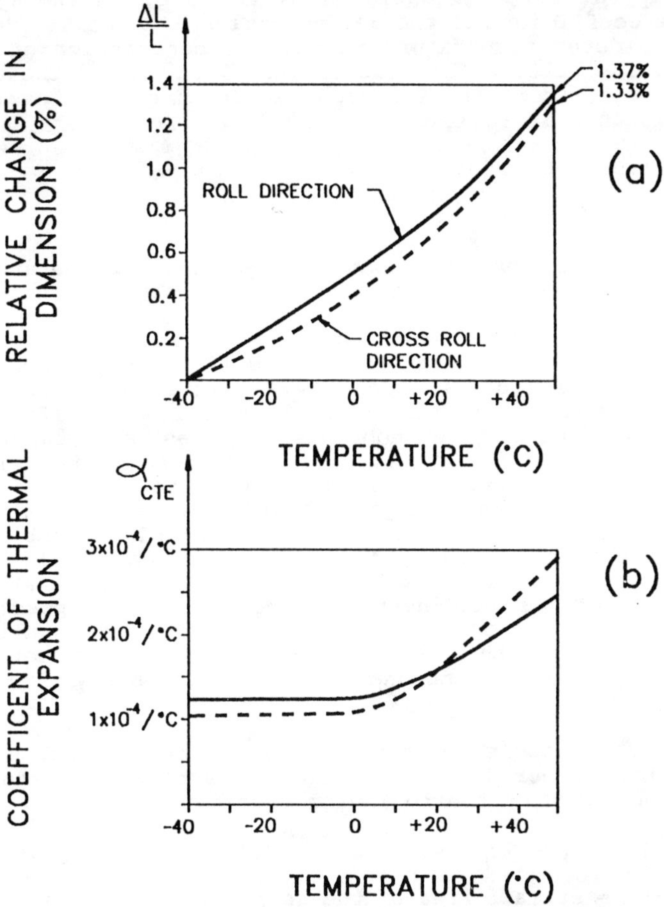

Figure 5. *Graphic Representation of Reversible Thermal Expansion-Contraction*: (a) curves giving the relative change in dimension using the length at -40°C as a basis; (b) curves of the coefficient of thermal expansion derived from the above curves. The geomembrane is an HDPE geomembrane, 2 mm thick, with the extrusion direction perpendicular to the roll direction. The specimens were 200 mm x 200 mm.

The required amount of slack can be calculated knowing the geomembrane coefficient of thermal expansion and the temperature differential to which it may be exposed. The worst case is when the geomembrane is at a high temperature during installation, e.g., 50°C, and is eventually subjected to a very low temperature, e.g.,-40°C. The temperature differential is then 90°C. Assuming an average coefficient of thermal expansion of $1.5 \times 10^{-4}/°C$ in the considered temperature range, the contraction of the geomembrane between the time of installation and the time it is subjected to the very low temperature is determined by Equation 5 as follows:

$$\Delta L/L = 1.5 \times 10^{-4} \times 90$$

$$= 1.4 \times 10^{-2} = 1.4\%$$

It should be noted that the geomembrane temperature, not the air temperature, is used in the calculation.

Installing geomembranes with slack is difficult with HDPE geomembranes because they are stiff. In particular, the intersection of wrinkles having different directions should be avoided. Installing with slack is also difficult with more flexible geomembranes such as PVC and CSPE geomembranes because the slack is displaced by the wind. Slack is more likely to remain in place with stiffer geomembranes such as HDPE geomembranes. Wrinkles that are intentionally placed on slopes to provide slack must have their axis along the slope (horizontal wrinkles would move downslope). To minimize the amount of required slack, geomembranes should be installed under the coolest possible temperature that is compatible with seaming quality. In the summer, this recommendation can by met by installing the geomembrane at night or early in the morning.

As a result of slack, the surface area of the geomembrane liner is larger than the surface area to be lined, as shown by the following equation:

$$A_G = A_L\ (1 + S_x + S_y) \qquad \text{(Equation 6)}$$

where: A_G = surface area of the geomembrane; A_L = surface area to be lined; S_X = required slack in the x direction; and S_y = required slack in the y direction.

For example, if the required slack is 1.4% in one direction and 1.6% in the other direction, the geomembrane should be 3% larger than the area to be lined. The impact on the cost of the project may be significant.

It should be noted that additional slack is required to alleviate the effect of irreversible shrinkage. As discussed in Section 3.4, irreversible shrinkage is mostly significant with PVC geomembranes. For example, if a PVC geomembrane requires a slack of 1.4% (calculated as shown above) to compensate for reversible thermal contraction and if a 2.5% irreversible shrinkage can be assumed per Section 3.4, the required amount of slack will be 3.9%.

3.6 Contraction of a Geomembrane Resting on Soil

In order to develop measures to be taken during installation to alleviate the influence of temperature on geomembrane behavior, it is necessary to understand the behavior of a geomembrane contracting or expanding, while it rests on the soil surface. Section 3.6 is related to a geomembrane contracting and Section 3.7 to a geomembrane exhibiting wrinkles as a result of thermal expansion. Calculations presented in these two sections are useful to evaluate the extent to which geomembranes resting on soil are free to move and to determine the required size and spacing of compensation panels used to remediate a geomembrane that was installed with insufficient slack.

If a geomembrane panel is placed flat on a horizontal soil surface and is subjected to a decrease in temperature, it tends to contract. However, interface friction between the geomembrane and soil tends to prevent contraction. If contraction cannot take place, tension is generated in the geomembrane. This tension is given by:

$$\alpha = E\ T_g\ \alpha_{CTE}\ \ \Delta T \qquad \text{(Equation 7)}$$

where: α = geomembrane tension; E = geomembrane modulus; α_{CTE} = geomembrane coefficient of thermal expansion; and ΔT = temperature decrease.

This tension is a maximum in the middle of a geomembrane panel (assuming the panel periphery is free, i.e., not connected to a structure or another panel) and it is balanced by the shear forces generated by friction, which are:

$$\alpha_F = \rho_g\ g\ T_g\ (L/2)\ \tan\phi_{sg} \qquad \text{(Equation 8)}$$

where: α_t = force per unit width resulting from shear stresses generated by friction; ρ_g = geomembrane density; g = acceleration of gravity; T_g = geomembrane thickness; L = length of geomembrane panel in the considered direction; and ϕ_{sg} = soil-geomembrane interface friction angle.

Equating the geomembrane tension with the force per unit width resulting from friction yields the following relationship that gives the amount of temperature decrease that is required to cause geomembrane contraction in spite of soil-geomembrane friction:

$$\Delta T = \rho_g \; g \; L \; \tan\phi_{sg} / (2 \; E \; \alpha_{CTE}) \qquad \text{(Equation 9)}$$

Calculations performed with this equation for a 30 m long geomembrane panel show that a temperature decrease of 2°C is sufficient to cause contraction of a high-modulus geomembrane, such as an HDPE geomembrane, whereas a decrease of 100°C is necessary to cause contraction of a low-modulus geomembrane, such as a PVC geomembrane. It appears that PVC geomembranes do not move as freely as HDPE geomembranes. However, even HDPE geomembrane panels do not move as freely as the above calculations would suggest, because an installed panel is usually either anchored in a trench or seamed to previously installed panels.

In Equations 8 and 9, the normal stress that mobilizes geomembrane-soil friction is the weight of the geomembrane. If the geomembrane is subjected to water pressure, Equation 8 becomes:

$$L = 2 \; E \; T_g \; \alpha_{CTE} \; \Delta T / (\rho_w \; g \; h_w \; \tan\phi_{sg}) \qquad \text{(Equation 10)}$$

where: ρ_w = density of water; h_w = depth of water; and other notations are as defined above.

Calculations performed with Equation 10 show that a depth of water of 1 m would be sufficient to prevent an HDPE geomembrane from moving between wrinkles that are 1 m apart even under a temperature difference of 30°C (which is highly unlikely under water). This is consistent with observations which have been made very often under water: a geomembrane subjected to tension for any reason, such as a differential settlement, does not move even in the vicinity of a wrinkle that could help alleviate the tension.

3.7 Wrinkling of a Geomembrane Resting on Soil

When a geomembrane is heated, it tends to expand. However, if it is laid flat on the soil, it cannot freely expand because of geomembrane-soil interface friction. As a result, wrinkles will develop. At a given temperature, the forces applied on a wrinkle include:

- lateral forces resulting from friction on the portion of geomembrane that remains in contact with the supporting soil;
- gravity forces on the wrinkle; and
- stresses in the wrinkle due to the stiffness of the geomembrane.

A theoretical analysis of this phenomenon has been conducted (Giroud and Morel, 1991) and has led to relationships giving the height of wrinkles and the distance between wrinkles as a function of the following parameters:

- the geomembrane expansion (resulting from temperature increase and geomembrane coefficient of thermal expansion);
- the geomembrane density;
- the geomembrane stiffness (resulting from the geomembrane thickness and modulus); and
- the interface friction angle between the geomembrane and the underlying soil.

A parametric study based on these equations shows that, for a given geomembrane expansion, the height of wrinkles and the distance separating wrinkles increase with increasing geomembrane stiffness and decreasing geomembrane-soil interface friction. Typical values obtained in this theoretical study, for a temperature increase of 50°C, are:

- a wrinkle height on the order of 0.1 m and a distance between adjacent wrinkles on the order of 10 m for HDPE geomembranes; and
- a wrinkle height on the order of 0.01 m and a distance between adjacent wrinkles on the order of 1 m for PVC geomembranes.

The parametric study also shows that the greater the geomembrane expansion, the higher and/or the closer the wrinkles for a given geomembrane. For example in one considered case with an HDPE geomembrane, a 0.6% expansion caused 70 mm high wrinkles located 16 m apart and a 1.2% expansion caused 90 mm wrinkles located 10 m apart.

In the field, wrinkles may move for the following reasons:

- Horizontal wrinkles that develop on a slope tend to move downslope as a result of gravity forces.
- The wind tends to displace wrinkles of low-modulus geomembranes (such as PVC geomembranes).

Observations show that the location of wrinkles that are parallel to geomembrane seams is often affected by the panel geometry: with some geomembranes, wrinkles occur systematically at seams, and with geomembranes that have manufacturing folds along the length, wrinkles tend to occur at these folds.

Observations made on wrinkles that have not been displaced by the wind and that are perpendicular to the seam direction in geomembranes laid on a horizontal soil surface are consistent with the theoretical results presented above. It is typically observed that HDPE wrinkles are large and that the distance between adjacent wrinkles is on the order of 10 m, whereas wrinkles are much smaller and closer in the case of PVC geomembranes.

3.8 Differential Expansion of Adjacent Panels

In the past two sections, it has been shown that geomembrane panels in contact with the supporting soil do not contract and expand freely. This may cause differential expansion of adjacent panels as explained below, using an example:

- On Day 1 in late afternoon, Panel 1 is installed. Its temperature is 60°C as a result of exposure to the sun.

- On Day 2 in early morning, Panel 2 is installed. Its temperature is equal to the ambient temperature, which is at its minimum, 15°C. Panel 1 has contracted during the night as a result of the decrease in geomembrane temperature. However, the contraction of Panel 1 has not been totally free and the actual contraction of Panel 1 is only 2/3 of the contraction that Panel 1 would have undergone if it had been totally free. Panel 1 is flat and under tension and Panel 2 is flat with no tension. The tension in Panel 1 corresponds to the constrained contraction due to a decrease of temperature of 15°C (i.e., 1/3 of the actual decrease of 45°C).

- On Day 2 at mid-morning the installer's crew is ready to seam Panel 1 and 2 together. The geomembrane temperature is then 25°C. The 10°C temperature increase (from 15°C to 25°C) is less than one third of the 45°C temperature decrease that occurred during the night between 60°C and 15°C. This 10°C temperature increase results in a partial tension release, but no wrinkles, in Panel 1 (which is still under light tension), and results in expansion and therefore wrinkles in Panel 2.

In the above example, the wrinkles at the edge of Panel 2, overlaying Panel 1 which is flat, form "fishmouths", which hamper seaming.

It should be noted that there would have been no fishmouths if Panel 2 had been installed when Panel 1 had reached the temperature where all its tension is released. (In the example presented above, this temperature would have been 30°C, which corresponds to one third of the temperature difference between 15° and 60°.)

3.9 Practical Recommendations

In various parts of Section 3, recommendations were made regarding measures that can be taken during geomembrane installation to alleviate the influence of temperature on the behavior of geomembranes. These measures are believed to have a beneficial effect on the quality and performance of installed geomembranes. They are recapitulated below to provide practical guidance to geomembrane installers and quality assurance monitors. Three types of measures are addressed successively: selection of time of installation; evaluation of required slack; and proper placement of slack. The following measures are mostly recommended for exposed geomembranes, which is often the case in liquid impoundments.

The optimum time for geomembrane installation can be selected taking into account the following considerations:

- In order to minimize thermal contraction, the geomembrane should be installed under the coolest possible temperature that is compatible with seam quality. In the summer, this condition can be met by seaming at night or early in the morning.

- In order to minimize the risk of differential wrinkles (known as fishmouths) between panels, no panel should be installed next to a panel under tension. Strategies similar to the one indicated in Section 3.8 should be developed on a case-by-case basis by the design engineer to prevent this situation from occurring. As a general rule, installation of panels early in the morning next to panels placed in the afternoon of the day before should be avoided.

The required geomembrane slack can be evaluated as follows:

- The required slack is the sum of slack components required to overcome geomembrane contraction due to loss of volatile components, stress relief, and thermal contraction.

- The slack required to overcome the loss of volatile components is negligible for HDPE geomembranes and can be evaluated using Equation 4 for PVC geomembranes. (This equation shows that the required slack for exposed PVC geomembranes will typically be between 1 and 3% assuming a plasticizer loss of between 10 and 25%.)

- The slack required to overcome stress relief cannot be precisely evaluated due to lack of data. Values of 1% for HDPE and 2% to 4% for PVC are suggested in the roll direction. Smaller values can be considered in the cross-roll direction.

- The slack required to overcome thermal contraction can be evaluated using Equation 5.

Slack should be placed in such a way that it will be effective. The following recommendations can be made:

- From the amount of required slack, the design engineer should prepare installation specifications regarding the size and spacing of the wrinkles. Small wrinkles (e.g., two centimeters high) closely spaced are recommended with PVC geomembranes, whereas any size and spacing is acceptable with HDPE geomembranes provided the wrinkles will not fold over under hydrostatic pressure.

- In the case of PVC geomembranes, sandbags should be placed every 10 m to prevent wrinkles from being excessively displaced by the wind.

- Wrinkles placed on slopes must have their axis along the slope (horizontal wrinkles would move downslope).

- After each panel is installed with wrinkles, it should be verified that the length and width occupied by the panel are shorter than the actual length and width of the panel by a percentage equal to the required amount of slack.

Some of the above recommendations are not appropriate for geomembranes that are not exposed (i.e., covered with a layer of protective soil and, in the case of landfills, with waste) because temperature variations, after covering, are much smaller. Less slack is needed in the case of geomembranes that are covered because all effects of temperature (loss of volatile components, stress relief, and thermal contraction) are reduced. It is fortunate that less slack is needed with geomembranes that are covered because, in this case, wrinkles can have many detrimental effects as discussed in Section 3.1. Finally, if a layer of soil is to be placed on a geomembrane that has been installed with slack, this should be done at the coolest time of the day, when the geomembrane is the flattest.

ACKNOWLEDGEMENTS

The authors are indebted to T.C. Bradford and N. Morel for their assistance during the preparation of this paper, to D.L. Pallanck and G. Saunders who typed the paper, K. Saimra and S. Kramer for drafting, and J.R. Mollenauer for editing. The authors are also grateful to J.E. Fluet, Jr., R. Bonaparte and K. Badu-Tweneboah for their review.

REFERENCES

Fluet, J.E., Jr., "Quality Assurance of Hazardous Waste Geosynthetic Liner Systems", *Proceedings of the Second Canadian Symposium on Geotextiles and Geomembranes*, Edmonton, Alberta, Canada, Sept. 1985, pp. 147-151.

Fluet, J.E., Jr., "Geosynthetic Lining Systems and Quality Assurance: State of Practice and State of the Art", *Proceedings of Geosynthetics '87*, Vol. 2, New Orleans, Louisiana, USA, Feb. 1987, pp. 530-541.

Fluet, J.E., "Quality Assurance: Planning Your Work and Working your Plan", *Geosynthetics World*, Vol. 1, No. 1, May 1990, pp. 20-25.

Giroud, J.P., and Kastman, K.H., "Guide specifications for construction of flexible membrane liners for hazardous waste disposal facilities", Vol. 1, pp. 4-67 to 4-99, in *Assessment of Technology for Construction and Installation of Cover and Bottom Liner Systems for Hazardous Waste Facilities* by M. Ghassemi, T.R.W., Inc., USEPA Contract No. 68-02-3174, 1983, 423 p.

Giroud, J.P., "Geomembrane liners: accidents and preventive measures", *Proceedings of the International R.I.L.E.M. Symposium on Plastic and Rubber Waterproofing in Civil Engineering*, Session 4, Liège, Belgium, June 1984a, pp. 1-6.

Giroud, J.P. "Opening Address", *Proc. of the International*

Conference on Geomembranes, Denver, Colorado, USA, Vol. 1, June 1984b, pp. 1-2.

Giroud, J.P. "Ageing of PVC Geomembranes in Uranium Mine Tailing Ponds", *Proceeding of the International Conference on Geomembranes*, Vol. 2, Denver, Colorado, June 1984c, pp. 311-316.

Giroud, J.P., "Analysis of Stresses and Elongations in Geomembranes", *Proceedings of the International Conference on Geomembranes*, Vol. 2, Denver, Colorado, USA, June 1984d, pp. 481-486.

Giroud, J.P. and Fluet, J.E., Jr., "Quality Assurance of Geosynthetic Lining Systems", *Geotextiles and Geomembranes*, Vol. 3, No. 4, 1986, pp. 249-287.

Giroud, J.P. and Bonaparte, R., "Leakage through Liners Constructed with Geomembranes, Part I: Geomembrane Liners", *Geotextiles and Geomembranes*, Vol. 8, No. 1, 1989, pp. 27-67.

Giroud, J.P., and Morel, N., "Analysis of Geomembrane Wrinkles", *Geotextiles and Geomembranes*, to be published, 1991.

Koerner, R.M., *"Designing with Geosynthetics"*, 2nd Edition, Prentice Hall, 1989, 652 p.

Koerner, G.R., and Bove, J.A., "Inspection of HDPE Geomembrane Installations", *Proceedings of Geosynthetics '89*, Vol. 1, San Diego, California, USA, Feb. 1989, pp. 70-83.

Matrecon, *"Lining of Waste Containment and Other Impoundment Facilities"*, USEPA, 600/2-88/052, Sept. 1988, 750 p. (plus Appendices).

Monteleone, M.J., and DiPippo, G.J., "Design Methods and Construction Quality Assurance For a Double Geocomposite Industrial Waste Landfill", *Proceedings of Geosynthetics '89*, Vol. 1, San Diego, California, USA, Feb. 1989, pp. 58-69.

National Sanitation Foundation, *"Standard Number 54 - Flexible Membrane Liners"*, Nov. 1985, Ann Arbor, Michigan, 26 p. (plus Appendices).

Peggs, I.D., "Evaluating Polyethylene Geomembrane Seams", *Proceedings of Geosynthetics '87 Conference*, Vol. 2, New Orleans, Louisiana, 1987, pp. 505-518.

Peggs, I.D. and Carlson, D.S., "Stress Cracking of Polyethylene Geomembranes: Field Experiences," *Durability and Aging of Geosynthetics* (Ed., R.M. Koerner), Elsevier Applied Science, London, UK, 1989, pp. 195 - 211.

Peggs, I.D., Carlson, D.S., Peggs, S.J., "Understanding and Preventing Shattering Failures of Polyethylene Geomembranes," *Proceedings of the Fourth International Conference on Geotextiles, Geomembranes and Related Products*, The Hague, The Netherlands, May 1990a, pp. 549-553.

Peggs, I.D., "Destructive Testing of Polyethylene Geomembrane Seams", *Geotextiles and Geomembranes*, Vol. 9, No. 4-6, 1990b, pp. 405-414.

Segrestin, P., and Jailloux, J.M., "Temperature in Soils and its Effect on the Ageing of Synthetic Materials", *Geotextiles and Geomembranes*, Vol. 7, No. 1-2, 1988, pp. 51-69.

USEPA, *"Minimum Technology Guidance on Double Liner Systems for Landfills and Surface Impoundments-Design, Construction and Operation"*, US Environmental Protection Agency, Washington, DC, USA, 1985, 71 p.

USEPA, *"Construction Quality Assurance for Hazardous Waste Land Disposal Facilities"*, US Environmental Protection Agency, 530-SW-86-031, Washington, DC, USA, October 1986, 88 p.

USEPA, *"Background Document: Proposed Liner and Leak Detection Rule"*, EPA/530-SW-87-015, May 1987, 526 p.

USEPA, *"Use of Construction Quality Assurance Programs and Control of Stress Cracking in Flexible Membrane Liner Seams"*, US Environmental Protection Agency, Memorandum, Washington, DC, USA, July 1989, 14 p.

Wallace, R.B., and Steinle, E.R., "Construction Quality Assurance of Geosynthetics Installations", *Proceedings of the 40th Canadian Geotechnical Conference*, Regina, Canada, Oct. 1987, pp. 365-373.

Wallace, R.B., "The Benefits of Construction Quality Assurance of Lining Systems Installation: Real or Perceived?", *Proceedings of Geosynthetics '89*, Vol. 1, San Diego, California, USA, Feb. 1989, pp. 84-94.

FIELD VERIFICATION OF CLAY LINER HYDRAULIC CONDUCTIVITY

By Gary W. Johnson[1], William S. Crumley[2], and Gordon P. Boutwell[3]

ABSTRACT

Test data from actual clay liner construction were analyzed to determine the correlation between field and laboratory measured values of hydraulic conductivity. A multi-variate regression analysis was also performed on data from four previously constructed clay liners to develop an equation to predict hydraulic conductivity as a function of the liquid limit, saturation, and density of the compacted clay. Two prototype clay liners were built using construction procedures similar to those used in cell liner construction. Pan lysimeters, sealed double-ring infiltrometers, and Boutwell permeameters were used to measure field conductivity values. Flexible wall and fixed-ring laboratory tests were used to measure laboratory conductivity values. Ratios of field to laboratory measured values ranged from 0.6 to 2.0. All field measured values were within a factor of 3.3 of the values predicted by the regression equation developed for this site.

INTRODUCTION

Solid and hazardous waste landfill owners and operators are being required by State and Federal regulatory agencies to demonstrate that clay liners have an in situ hydraulic conductivity less than 1×10^{-7} cm/sec. Day and Daniel (1985) reported results of field and laboratory tests showing a large variation between field and laboratory hydraulic conductivity values. Their tests indicated that ignoring basic principles for constructing low permeability clay liners, resulted in a clay liner with high hydraulic conductivity. As

---

[1]Manager, Geotechnical Engineering; Browning-Ferris Industries; Houston, Texas
[2]Regional Engineer; Browning-Ferris Industries; Atlanta, Georgia
[3]President; Soil Testing Engineers, Inc.; Baton Rouge, Louisiana

pointed out by 19 discussers, these liners were not built using sound principles for constructing low permeability liners. Consequently, owners and operators are being asked to perform in situ hydraulic conductivity tests on prototype liners and even on the actual clay liner.

The technical staff of Browning-Ferris Industries, Inc. (Workman and Oakley, 1987) believed that liners constructed using fundamentally sound criteria would have hydraulic conductivities close to the measured laboratory values. Because of this, BFI initiated a testing program at one of its waste disposal sites to show that properly constructed liners would indeed have low hydraulic conductivity. The testing program involved: (1) constructing a prototype liner with lysimeters beneath the liner to measure the breakthrough time and flow rates of impounded fluids, and, (2) performing a field hydraulic conductivity test on a separate prototype liner using a sealed double-ring infiltrometer (Daniel and Trautwein, 1987). Both liners were constructed using material, equipment, and procedures consistent with construction specifications normally used at the facility. In addition, Boutwell field permeameters (Daniel, 1989) were used to provide a third measure of in situ hydraulic conductivity of one of the prototype clay liners. The site was chosen because of the large database of field and laboratory quality control testing from the previously constructed disposal cells and impoundments.

PURPOSE AND SCOPE

The primary purpose of this research was to show that compacting clay using procedures consistent with waste industry standards would produce a liner with in situ hydraulic conductivity comparable to the conductivity measured by laboratory tests on undisturbed samples from the liner. Another purpose was to develop a statistical relationship between hydraulic conductivity tests and liner density, saturation, and liquid limit for the clay liners constructed at this facility.

The scope of work included analyzing all laboratory data from tests performed on the four most recently constructed disposal cell liners to develop the statistical relationship. These laboratory test results, as well as test results on samples from the two protype liners, were then compared to the hydraulic conductivities calculated from the field tests to see how close laboratory values compared to the in situ measurements. All measured results were compared with the predicted results to validate the predictive equation.

CLAY LINER BORROW MATERIAL

Clay liners for 22 cells and impoundments have been constructed at this facility during the past 12 years. The clay liner materials have been taken from on-site excavations and from off-site borrow pits in the same geological formation: The Prairie Terrace Complex (or "Beaumont Clay"), a Pleistocene-age deposit. These soils have been classified as CH and high-plasticity CL materials according to ASTM D2487. The liquid limits (w_L) for the materials of the individual cells ranges from about 42 to 68. The sand contents of these soils are typically less than 10%.

Four disposal cells were built between 1986 and 1990. Data on the plasticity characteristics of the liner soils from these cells, plus Prototype Liners A and B are presented below:

TABLE 1 - LINER PLASTICITY

Unit (1)	No. of Tests (2)	Mean w_L (%) (3)	Dev. w_L (%) (4)	Mean Ip (%) (5)	Dev. Ip (%) (6)
Liners I-V	377	45.6	7.2	29.5	7.0
Prototype Liner A	8	49.8	3.5	34.4	3.0
Prototype Liner B	9	50.0	3.3	34.4	2.8

The compaction characteristics of the materials were determined using the ASTM D698 (Standard Proctor) method. The maximum dry densities (D_m) ranged from 94 to 114 lbs./cu.ft. (1.51 to 1.83 g/cc) and the optimum moisture contents (w_o) from 15 to 25%. As expected, materials with higher (w_L) values tended to have higher optimum moisture contents (w_o) and lower dry densities (Dm). A linear regression on the 54 Standard Proctor tests resulted in the following relationships:

$$D_m = 123.4 - 0.327\, w_L \quad \text{......................(1)}$$
$$w_o = 10.9 + 0.140\, w_L \quad \text{......................(2)}$$

The Correlation Coefficients (Cc) were 0.71 and 0.81, with Standard Estimates of Error (SEE) of 2.5 pcf (0.04

gm/cc) on dry density and 1.5 percentage points on optimum moisture contents.

Laboratory hydraulic conductivity tests were performed on samples compacted wet of optimum in the laboratory as a materials quality assurance procedure. These were conducted using fixed-wall permeameters and hydraulic gradients near 50. The 33 tests had a mean w_L of 51, compaction density of 95% of maximum Standard Proctor, and saturation(s) of 0.90 (3.9 percentage points over w_o). The mean geometric permeability of this test group was 6 x 10^{-9} cm/sec.

CLAY LINER HISTORICAL DATA

The clay liners were constructed under rigid Construction Quality Assurance (CQA). The specifications for the most recent four liners are typical:

1. Liquid Limit: 30% (minimum)
2. Sand Content: 50% (maximum)
3. Compaction Density: 90% of Standard Proctor (ASTM D698) (minimum)
4. Moisture Content: Optimum + 2% to 8%
5. Hydraulic Conductivity: 1 x 10^{-7} cm/sec (maximum)

The CQA plan for these cells included field moisture-density testing for preliminary acceptance under criteria (3) and (4), followed by laboratory permeability tests (fixed-ring samples with gradient of 50) on undisturbed samples taken from the compacted liner. Some pertinent data from the CQA programs on these four clay liners are presented in Table 2.

TABLE 2 - LINER CQA DATA

	Liquid Limit		Compacted Density*		Saturation		Hydraulic Conductivity (cm/sec)**	
Liner	Tests	Mean	Tests	Mean	Tests	Mean	Tests	Mean
(1)	(2)	(3)	(4)	(5)	(7)	(8)	(9)	(10)
I	64	42	507	94	507	0.88	64	4 x 10^{-9}
II	86	50	660	93	660	0.90	90	5 x 10^{-9}
III	82	47	411	94	411	0.93	82	3 x 10^{-9}
IV	152	44	759	93	759	0.85	141	4 x 10^{-9}

* As percent of D_m from ASTM D698.

Hydraulic conductivity (or permeability) is log-normally distributed, i.e., permeability itself does not follow a Normal Distribution, but the logarithm of permeability does (Bogardi, 1987; Boutwell and Hedges, 1989). Figure 1 is a histogram showing the logarithmic distribution of the 377 permeability tests on liners for Cells I through IV, compared with the appropriate (log) Normal Distribution curve. A close fit is apparent.

It is commonly accepted that a decrease in permeability (k) results from an increase in clay content or activity, density, or saturation at compaction. The 377-test database was subjected to multivariate least-squares regression analysis using three variables following Boutwell and Hedges (1989). The best-fit equation was:

$$\text{Log}(k) = -(1.8472 + 1.986\text{Log}(w_L) + 1.159(S^3) + 0.0248(D))\ldots.(3)$$

where (S) is the saturation at compaction and (D) is dry density (in pcf), both measured in the field. The degree of fitting is shown on Figure 2, which illustrates values of hydraulic conductivity predicted from Equation (3), and the results from laboratory tests on samples obtained from the test liners. The laboratory data fall within the general range of the database tests. The mean ratio of measured to predicted values for the database is about 1.0 (as it must be). It should be noted that for about 80 percent of the data used to generate Equation 3, the predicted value is within a factor of three of the measured value.

PROTOTYPE LINER CONSTRUCTION

In situ tests were performed on two different prototype liner sections constructed on-site. The liner sections were constructed using the clay borrow material described above and a Caterpillar 815 compactor. The compactor has an approximate weight of 44,000 lbs (20,000 kg) and applies a non-vibratory force of about 3440 lbs per ft. (477 kg per meter) of drum width to the clay liner.

Prototype Liner A was designed to duplicate, (as closely as possible on a large scale) the boundary and flow conditions of a fixed-wall, constant head laboratory permeameter using actual construction equipment for compaction. Important considerations included certainty of vertical flow lines, constant

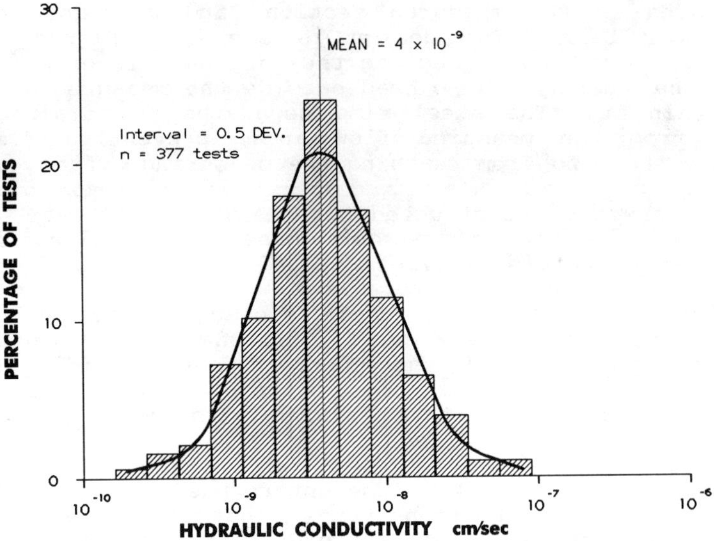

FIG. 1. Histogram of logarithmic distribution of permeability tests on Liners I,II,III, and IV.

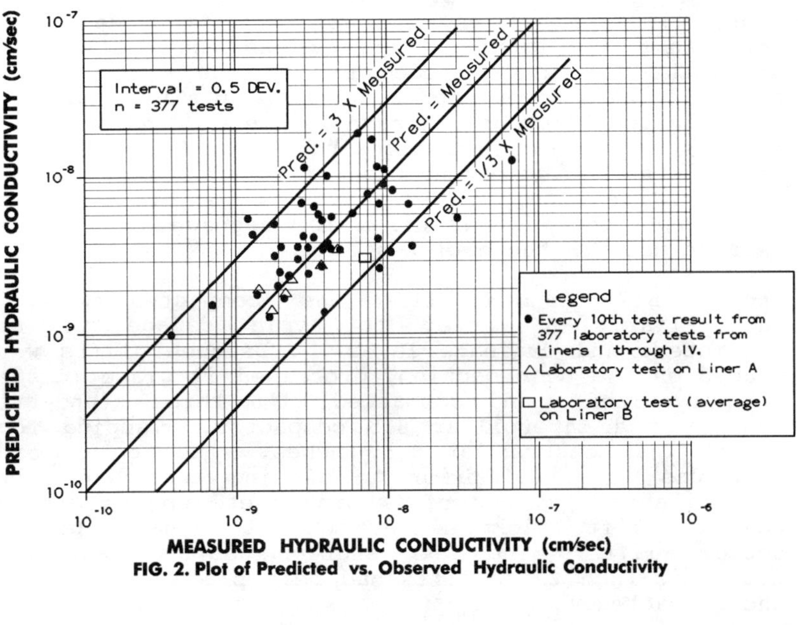

FIG. 2. Plot of Predicted vs. Observed Hydraulic Conductivity

flow area in the measured section, and constancy of head over time. The design is shown in Figure 3. Vertical flow occurs because the drainage layer maintains the same hydraulic head outside the measured area as within it. The steel ring separates the drainage layer from the measured flow area, preventing flow outside the ring from reaching the measuring system.

The liner was constructed in early 1986 over a compacted one foot thick soil base, into which the concrete collectors (lysimeters) and outlet piping were installed. Eight lysimeters were installed, as shown in Figure 3, and the concrete sloped to the outlet pipe. When the concrete cured, silicone caulk was used to seal the edge of the concrete both inside and outside the steel ring. The concrete surfaces inside the steel rings were then painted with two coats of epoxy sealer.

The drainage layer covered the entire base area outside the steel rings and was butted against the rings. In preparation for the first lift of the test section, the outlet pipes were extended outside the base area and up to the elevation of the top of the concrete just inside their respective rings. Inside the rings, a woven fabric was placed on the concrete, followed by a nonwoven filter fabric. Then a clean sand was added to bring the collector surface level. The sand was then covered with a woven fabric. This maintains a constant test section thickness at each lysimeter, and prevented the test fill clay from blocking the outlet pipe flow. Lastly, the outlet piping was filled with water, to the level of the sand, to aid in saturation of the system and lessen the time required for seepage in the test section to fill the outlet piping.

The first six-inch lift was compacted with a Caterpillar D-6D dozer to avoid damage to the lysimeters and drainage layer. Subsequent lifts were placed on the test section, processed in place (disced or pulverized), and compacted with three round-trip passes of a Caterpillar 815 compactor. Tamping foot length was measured at 8.5 inches. The top of each lift was scarified prior to placing the next lift. Scarification and use of tamping feet lengths greater than the lift thickness were used to enhance bonding between lifts. The test section was finished when four, six-inch thick lifts had been placed, processed, and compacted.

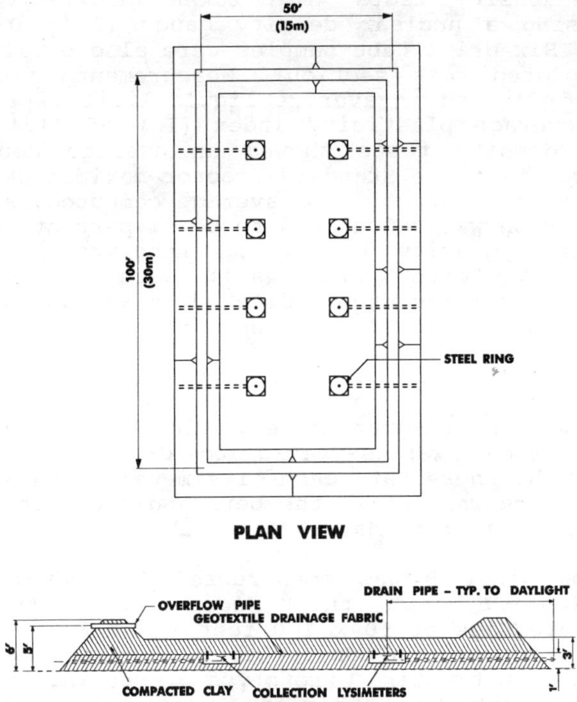

FIG. 3 Prototype Liner With Collection Lysimeters

During test section construction, clay moisture content was checked frequently and verified to be from 0 to 6 percent wet of optimum (ASTM D698). Maximum clod size was 2 to 4 inches. After processing, the clods appeared to be uniform in water content. Moisture-density tests were taken of the compacted lifts using a nuclear density gauge (ASTM D3017 and D2922). Six drive tube samples were also obtained from the completed test section. Measurements from field samples indicated an average liquid limit (w_L) of 49.8 and an average plasticity index (I_p) of 34.4. Field moisture-density tests showed an average density of 95.8 percent of the Standard Proctor maximum density or 104.4 pcf (1.67 g/cc). The average computed saturation at compaction was calculated to be 91 percent, assuming a specific gravity of 2.72. Laboratory hydraulic conductivity tests (fixed-wall) were run on seven undisturbed samples, and indicated an average hydraulic conductivity of 3×10^{-9} cm/sec.

Following completion of the test lifts, a berm was constructed outside the lysimeter area and the fill surface was smooth-drum rolled. The fill surface above each lysimeter was surveyed to determine the test section thickness at each lysimeter location. An overflow pipe was set in the berm and the test fill was filled with water in May 1986.

Prototype Liner B was constructed in September 1987 using four 6 inch thick compacted lifts to an approximate size of 50 x 100 feet (15 x 30m). Similar methods were used in constructing this section as were used in constructing Prototype Liner A, with the exception that the lysimeters were not constructed beneath the clay liner, and the soil was not pulverized, only disced during processing. A non-woven geotextile was laid below the liner to provide a drainage way during testing. Samples taken from the liner section indicate an average liquid limit (w_L) of 50.0 and a plasticity index (I_p) of 34.4. The field density tests indicated an average density of 93.5% of maximum density (ASTM D698) or 101.9 pcf (1.63 g/cc). The average computed saturation at compaction was 88 percent, assuming a specific gravity of 2.72.

IN SITU PERMEABILITY TESTS

Prototype Liner A

This liner had eight lysimeters below the clay and a berm around the liner to contain water that was ponded on the liner. The berm allowed infiltration of water through the liner under a hydraulic head of approximately 2 ft. (0.61 m). By monitoring the time until water shows in the lysimeters, it was hoped that the saturated hydraulic conductivity could be calculated. Also, by ultimately measuring the flow rate once steady - state flow was achieved, the hydraulic conductivity could be calculated.

The eight lysimeters were monitored beginning on May 6, 1986 when the pond was filled with water, to determine when water started moving through the clay liner. Fluid began showing in two of the collectors in January 1989, 32 months after filling the pond. The volume of water collected was erratic, varying from 0 to 100 cm^3/day. As of March 1990, water was showing in four of the eight collectors, with volumes ranging from 1 to 54 cm^3/day. Since water began showing, the average flow per day for the four collectors is 11.9 cm^3/day with a range between collectors of 6.8 to 21 cm^3/day.

Green and Ampt (1911) proposed a solution that can be used to calculate the hydraulic conductivity of a soil, based upon the breakthrough time, from the following equation:

$$k = (w_v/t)[L - (h + P)\ln(1 + L/(h + P))] \quad \text{....(4)}$$

where k = saturated hydraulic conductivity,
w_v = change in volumetric water content,
t = time to saturation of the liner or breakthrough time,
L = total liner thickness,
h = head of water on liner, and
P = absolute value of soil suction at wetting front.

This equation is based upon the simplifying assumption that the wetting front moves downward through the liner at a uniform rate, with a distinct boundary between saturated and unsaturated soil.

Using Equation 4, hydraulic conductivity can be calculated using the following values:

w_v = 0.04 (calculated using an initial saturation of 91%)
t = 32 months = 84,096,000 sec.
L = 2 ft. = 61 cm
h = 2 ft. = 61 cm
P = unknown

Many authors (Bouwer, 1964; Bouwer, 1969; Brakensiek, 1977) have presented methods for estimating the suction to be used in calculating hydraulic conductivity from the Green-Ampt model. Bouwer (1978) suggested that the suction head at the wetting front may be estimated by using the water-entry (or air-exit) suction value. This water-entry suction value may be estimated as being one-half of the air-entry suction value (bubbling pressure or capillary rise) according to Bouwer (1969 and 1978). A set of laboratory suction (capillary rise) tests were run on three samples compacted to 95% of maximum density (ASTM D698). The test results were as follows:

TABLE 3 - SATURATION VS. SUCTION TEST DATA

Saturation, %	Capillary Rise cb	cm of H_2O
(1)	(2)	(3)
83	60	621
90	34	352
94	28	290

The water-entry suction value that should be used in the Green-Ampt model is not the initial suction value in the clay liner, but a weighted value to account for the change in soil suction with a corresponding change in volumetric water content. Brakensiek (1977) concluded that using 76 percent of the bubbling pressure (or air-entry suction) value was appropriate, but it was not clear if the initial value or weighted value of air-entry suction should be used.

For the purpose of this analysis, we assumed that the initial value of air-entry suction was 330 cm of water (interpolated from Table 3 for an initial saturation of 91 percent) and at full saturation the suction would be 0 cm. If it is assumed that the air-entry suction value decreases linearly as the wetting front advances, an average value of 165 cm may be used. In accordance with Bouwer (1969 and 1978), the water-entry suction value was estimated to be one-half the air-entry suction value or 82 cm of water. Averjanov, (1949), suggested that the appropriate value of water-entry suction should account for the decreasing hydraulic conductivity in the capillary rise zone by applying a factor of 0.3 to the initial capillary rise value. This would result in a suction value of 99 cm. Using the more conservative suction value of 82 cm in the Green-Ampt model results in a calculated hydraulic conductivity of 5 x 10^{-9} cm/s.

Hydraulic conductivity may also be calculated knowing the flow rate in the liner, using the Darcy equation for saturated flow as discussed by Daniel and Trautwein (1986):

$$k = Q/iA \quad \ldots\ldots\ldots\ldots\ldots\ldots\ldots\ldots\ldots\ldots (5)$$

where k = saturated hydraulic conductivity,
Q = measured outflow through lysimeter,
i = hydraulic gradient, and
A = area of lysimeter.

As discussed above, the flow through the lysimeters has been erratic, indicating steady - state flow and saturation has not yet been achieved.

To estimate permeability, the following parameters were used:

Q = 11.9 cm^3/day (average outflow value)
i = (L + H)/L = 2
A = 11,680 cm^2

Using the average outflow value, the Darcy equation results in a calculated hydraulic conductivity of 6 x 10^{-9} cm/sec. This is in good agreement with the permeability calculated using the Green-Ampt model above.

Prototype Liner B

This liner was constructed in September 1987. A Sealed Double-Ring Infiltrometer (SDRI) (Daniel and Trautwein, 1986) was constructed on this liner in October 1987. The SDRI consisted of a sealed inner ring (1.5 x 1.5 m) with an open outer ring (3.7 x 3.7 m) that were embedded into the clay liner and sealed with a bentonite grout. Both rings were filled with water to a height of 30 cm above the liner. A flexible bag filled with water was connected to the inner ring and placed in the water between the rings. The bag was periodically disconnected from the sealed inner ring and weighed to determine the incremental flow of water into the liner within the inner ring over a given time period. As discussed by Daniel and Trautwein (1986), the infiltration rate was used to calculate an in-situ hydraulic conductivity using the following equation:

$$k = I/i \quad \ldots\ldots\ldots\ldots\ldots\ldots\ldots\ldots (6)$$

where k = hydraulic conductivity, cm/sec,
I = measured infiltration, cm/sec, and
i = hydraulic gradient, cm/cm.

In calculating the hydraulic conductivity, the hydraulic gradient must be estimated. The gradient may be estimated using the following equation:

$$i = (H + L + P)/L \quad \ldots\ldots\ldots\ldots\ldots\ldots (7)$$

where H = height of water on liner, cm,
L = thickness of liner, cm, and
P = absolute value of soil suction, cm.

According to Daniel (1989), the suction is usually assumed to be zero in calculating the permeability. This can be conservative if suctions exist at the wetting front. Figure 4 shows a plot of infiltration

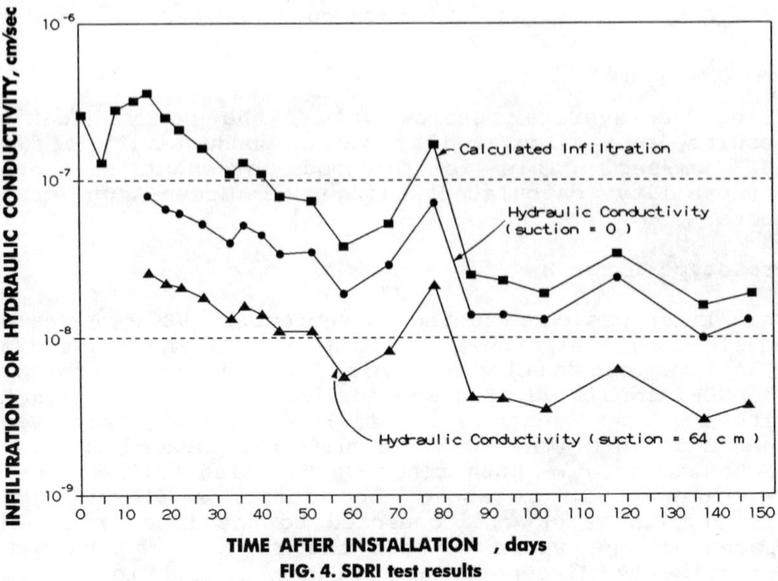

FIG. 4. SDRI test results

as measured during the SDRI test as well as hydraulic conductivity calculated using two different hydraulic gradients. In calculating the hydraulic gradient, it was first assumed that full depth saturation had occurred, thus P = 0 and i = 1.5. The authors felt this was conservative for two reasons: First, using the calculated void ratio and saturation, and knowing the quantity of flow into the liner, the average depth of uniformly advancing wetting front could be calculated. From this calculation, it was apparent that the average wetting front had only reached a depth of about 0.2 m through the 0.6 m thick liner. Second, the tensiometers at 0.45 m depth had an average reading of 300 cm, indicating full saturation had not been achieved below the wetting front. It is possible that the flow had found one or more preferential pathways through the liner and thus was not advancing uniformly, but it can also be argued that the infiltration rate was still dropping after 140 days, indicating that the gradient was still dropping and therefore, the liner had not reached full saturation. Therefore, a second hydraulic conductivity was calculated using the calculated uniform wetting front depth and a suction value. At the end of the test (140 days), the tensiometer readings (from all depths) averaged 129 cm of water. In accordance with Bouwer (1969 and 1972), this capillary rise value (or air-entry value) was reduced by 50 percent to estimate the water-entry suction value to be used. Therefore, a value of 64 cm of water was used in computing the second hydraulic conductivity. Figure 4 indicates that after only 40 days, the infiltration rate had dropped below 1×10^{-7} cm/sec and the final calculated hydraulic conductivities were in the range of 4×10^{-9} cm/s to 1×10^{-8} cm/s, more than one order of magnitude below the required 1×10^{-7} cm/sec.

The SDRI device was dismantled in March 1988. At this time, two thin-walled (7.6 cm diameter) tube samples were taken between the inner and outer ring. Four flexible-wall permeability tests were run on specimens from these samples under an effective confining pressure of 5 psi (35 kPa) and a gradient of 182 to 185. The range of hydraulic conductivity values from these laboratory tests was 3×10^{-9} cm/sec to 2×10^{-8} cm/sec with an average of 7×10^{-9}.

Boutwell Permeameter

Prototype Liner A was also investigated using the two-stage field permeability test originated by one of the authors (Boutwell and Derick, 1986; Daniel, 1989). This test allows direct, in-place determination of hydraulic conductivity in both the vertical and horizontal directions, in this case, normal (k_n) and parallel (k_p) to the plane of compaction. By varying the geometry of the open hole area during the test, the relative effects of the vertical and horizontal conductivities are varied. The test is conducted in a cased borehole. In the first stage, the bottom of the borehole is flush with the bottom of the casing. Vertical conductivity has its greatest effect on the test during this stage. The borehole is then advanced beyond the bottom of the casing, and the test repeated. The effect of horizontal conductivity is greater during this stage. The current data reduction method uses the Hvorslev (1951) equations for a homogeneous, anisotropic infinite medium. The two stages provide two Hvorslev equations with two unknowns (vertical and horizontal conductivity), so that a solution for these unknowns is possible. The liner was only 2 feet (60 cm) thick, so that the originator's recommendations of maintaining 5 diameters embedment below the top of liner and the same amount of remaining soil below the bottom of the test to approach an infinite medium could not be met. Hence, boundary effects were expected to affect data reduction.

Six such tests were set up; three were 2 inch (5 cm) diameter installations and three had 4 inch (10 cm) diameters. All were installed through the perimeter levee and began with an embedment of about 6 inches (15 cm) below the top of the actual liner. The first stage of each test achieved a steady-state condition in 6 to 7 days, but was maintained for about 30 days. At that point, the tests were extended for the second stage and undisturbed samples of the removed liner material were taken for laboratory testing. The second stage achieved steady-state in 5 to 6 days, and was also maintained for 30 days. The results of the field and laboratory programs are summarized in Table 4. The terms k_1 and k_2 are the direct results from Stages 1 and 2, respectively (isotropic medium assumed).

As suggested, the two-stage field test program was not conclusive due primarily to boundary effects. The current data reduction method assumes an infinite medium and this requires sufficient clearance (at <u>least</u> 5 diameters) above and below the test intake area. Furthermore, any variation in density (and thus saturation and permeability) through the thickness of a lift violates the Hvorslev assumption of a homogeneous though anisotropic medium for small-diameter tests. This is clear from the difference in results from the 2 and 4 inch (5 and 10 cm) diameter tests.

It is interesting to note that results quite similar to those from the CQA test samples and the lysimeters can be obtained empirically. Assuming the ratio (k_p/k_n) for compacted clay to lie in the range of 5 to 10 (Boutwell and Rauser, 1990), k_n can be computed from the Stage 1 results (k_1) which should be least affected by boundary conditions. The mean value for k_n computed from the 6 two-stage tests is then 4 to $6x10^{-9}$ cm/sec, while the results from laboratory tests on undisturbed samples from Liner A averaged $3x10^{-9}$ cm/sec and the lysimeter results ranged from 5 to $6x10^{-9}$ cm/sec.

TABLE 4 - TWO STAGE BOUTWELL PERMEAMETER RESULTS

Field Test

Item (1)	BP-1 (2)	BP-2 (3)	BP-3 (4)	BP-4 (5)	BP-5 (6)	BP-6 (7)
Diameter	2"	4"	2"	4"	2"	4"
Stage 2 L/d	None	1.4	2.4	1.1	2.2	1.5
k_1-(F)	$2x10^{-9}$	$2x10^{-8}$	$6x10^{-8}$	$8x10^{-9}$	$7x10^{-9}$	$6x10^{-8}$
k_2-(F)	Leak	$4x10^{-8}$	$3x10^{-8}$	$2x10^{-8}$	$4x10^{-8}$	$2x10^{-8}$
k_n-(F)	-	$3x10^{-9}$	-	$8x10^{-10}$	$6x10^{-10}$	-
k_n-(L)	$2x10^{-8}$	$9x10^{-9}$	$9x10^{-9}$	$6x10^{-9}$	$2x10^{-8}$	$7x10^{-9}$
k_p-(F)	-	$1x10^{-7}$	-	$8x10^{-8}$	$1x10^{-7}$	-
k_p-(L)	-	$1x10^{-8}$	-	$2x10^{-8}$	-	$5x10^{-9}$
k_p/k_n-(F)	-	38	<1	95	225	<1
k_p/k_n-(L)	-	13	-	29	-	-
w-%	28	24	24	25	-	27
D-pcf	103	93	110	95	-	95
w_L - %	-	47	-	52	-	52
I_p - %	-	31	-	34	-	35

F: Field determination
L: Laboratory determination (fixed-wall, gradient = 50)
L/d: Ratio of open hole depth to diameter
k_1: Direct result of Stage 1, (cm/s)
k_2: Direct result of Stage 2, (cm/s)
k_n: Permeability normal to plane of compaction, (cm/s)
k_p: Permeability parallel to plane of compaction, (cm/s)
w, D, w_L, and Ip: As previously defined

ANALYSIS AND DISCUSSION OF PERMEABILITY RESULTS

The laboratory and in situ hydraulic conductivity test results performed for this study are summarized in Table 5. The data indicates that using the moderately plastic clay and the construction procedures and equipment outlined above, a clay liner can be constructed that has a measured hydraulic conductivity substantially less than $1x10^{-7}$ cm/sec. The average hydraulic conductivity values determined by laboratory testing ranged from $3x10^{-9}$ cm/sec to $7x10^{-9}$ cm/sec. The three in situ tests on the two Prototype Liners ranged from $4x10^{-9}$ cm/sec to $1x10^{-8}$ cm/sec. For the Prototype Liners, the ratio of field to laboratory hydraulic conductivity (k_f/k_l) ranged from 0.6 to 2.0 as shown in Table 5. The highest calculated value of hydraulic conductivity from the in situ tests was $1x10^{-8}$ cm/sec. This is conservative because soil suction was assumed to be negligible when in fact the liner was not saturated.

TABLE 5 - SUMMARY OF RESULTS

Hydraulic Conductivity cm/sec

Test (1)	k_m (2)	k_p (3)	Saturation (4)	k_f/k_l (5)	k_m/k_p (6)
Liners I-IV (Lab)	$4x10^{-9}$	$4x10^{-9}$	0.88	-	1
Liner A (Lab)	$3x10^{-9}$	$2x10^{-9}$	0.91	-	1.5
Liner A (Green-Ampt)	$5x10^{-9}$	$2x10^{-9}$	0.91	1.7	2.5
Liner A (Darcy)	$6x10^{-9}$	$2x10^{-9}$	0.91	2	3
Liner A (Boutwell)*	$6x10^{-9}$	$2x10^{-9}$	0.91	2	3
Liner B (Lab)	$7x10^{-9}$	$3x10^{-9}$	0.88	-	2.3
Liner B (P = 0)	$1x10^{-8}$	$3x10^{-9}$	0.88	1.4	3.3
Liner B (P = 64 cm)	$4x10^{-9}$	$3x10^{-9}$	0.88	0.6	1.3

k_f: field hydraulic conductivity

k_l: laboratory hydraulic conductivity

k_m: measured hydraulic conductivity

k_p: predicted hydraulic conductivity from Eq. 3

* Calculated from first stage (vertical) assuming $k_p/k_n = 5$

Also shown in Table 5 are the ratios of measured to predicted hydraulic conductivity (k_m/k_p). The predicted values of hydraulic conductivity as developed from Equation (3) are based upon a statistical relationship using liquid limit, saturation, and dry density. The ratio of k_m/k_p ranged from 1.3 to 3.3 for the two Prototype Liners.

As discussed earlier, it is a common belief that compacted clay liners allow large volumes of liquids to pass through the clay. These results indicate otherwise. Reades, et al, (1986) and Gordon, et al, (1989) also present data from four other landfills that indicate good agreement between field and laboratory hydraulic conductivity data. It is important to point out that for those four cases as well as this study, the clay liners were constructed using fundamentally sound construction principles.

CONCLUSIONS

A study was performed on clay liners at a waste disposal site in Louisiana to compare laboratory and in situ hydraulic conductivity values. A secondary purpose was to establish a statistical relationship between laboratory hydraulic conductivity tests and liner density, saturation at compaction, and liquid limit for the clay liners.

The results indicated that laboratory hydraulic conductivity values generally agreed well with in situ values. The ratio of field to laboratory hydraulic conductivity ranged from 0.6 to 2.0 A statistical relationship was developed for predicting hydraulic conductivity values based upon clay sample indices. Field measurements of clay liner conductivity show good statistical correlation to the developed equation. All field measurements were within a factor of 3.3 of the predicted values.

The study indicates low permeability clay liners can be constructed using fundamentally sound construction principles; and the resulting hydraulic conductivity can reasonably be predicted using laboratory tests and statistical correlations.

ACKNOWLEDGEMENTS

The authors would like to thank Browning-Ferris Industries, and Soil Testing Engineers, Inc. for funding this study. The authors also acknowledge the contributions of Messrs. Robert McElligott, Stephen Trautwein, John Workman, Ray Wallace, Steve Yob, Daniel Spikula, Kareen Van Deman, and Bertrand S. Palmer; and Messrs. David E. Daniel, Charles A. Rivette, and Richard E. Oakley for reviewing the manuscript.

APPENDIX I: REFERENCES

Averjanov, S.F., "Approximate Appraisal of the Role of Seepage in the Zone of Capillary Fringe," Doklady Akademii Nauk, CCCP, Vol. 69, No. 3. 1949, pp. 309-311.

Bogardi, I., Kelly, W.E., and Bardossi, A (1987), "Component and Systems Reliability for Composite FML Containment Systems, Phase I Report." EPA Project CR813433-01, University of Nebraska, Lincoln, Nebraska, August 1987.

Boutwell, G.P., and Rauser, C.L. (1990), "Clay Liner Construction," Proc. of ASCE/Penn. DOT Geotech. Seminar, Hershey, Pa., April 1990.

Boutwell, G.P., and Derick, R.K. (1986), "Groundwater Protection for Sanitary Landfills in the Saturated Zone." Presented to Waste Tech 1986, NSWMA, Chicago, Ill.

Boutwell, G.P., and Hedges, C.S. (1989), "Evaluation of Waste-Retention Liners by Multivariate Statistics." Proc. XII Int. Conf. on Soil Mech. and Found. Eng., Vol. 2, Rio de Janiero, August 1989, pp. 815-818.

Bouwer, H. (1964), "Unsaturated Flow in Ground-Water Hydraulics," J. Hydraulics Div., ASCE, Vol. 90 No. HY5, September 1964, pp. 121-144.

Bouwer, H. (1969), "Infiltration of Water into Non-uniform Soil," J. Irrigation and Drainage Div., ASCE, Vol. 95, No. IR4, December 1969, pp. 451-462.

Bouwer, H. (1978), Groundwater Hydrology, McGraw-Hill, New York, N.Y.

Brakensiek, D.L. (1977), "Estimating the Effective Capillary Pressure in the Green and Ampt Infiltration Equation," Water Resources Research, Vol. 13, No. 3, June 1977, pp. 680-682.

Daniel, D.E. (1989), "In Situ Hydraulic Conductivity Tests for Compacted Clay," J. Geotech. Engrg., ASCE, Vol. 115, No. 9, September 1989, pp. 1205 - 1226.

Daniel, D.E. and Trautwein, S.J. (1986), "Field Permeability Test for Earthern Liners," Proc., In Situ 1986, ASCE, Blacksburg, Va., pp. 146-160.

Day, S.R. and Daniel, D.E. (1985), "Hydraulic Conductivity of Two Prototype Clay Liners." J. Geotech. Engrg., ASCE, Vol. III, No. 8, August 1985, pp. 957-970.

Green, W.H. and Ampt, G.A. (1911), "Studies in Soil Physics Part I - The flow of air and water through soils," J. Agric. Science, Vol. IV, pp.1-24.

Gordon, M.E., Huebner, P.M., and Miazga, T.J. (1989), "Hydraulic Conductivity of Three Landfill Clay Liners," J. Geotech. Engrg., ASCE, Vol. 115, No. 8, August, 1989, pp. 1148-1160.

Hvorslev, M.J. (1951), Time lag in the observation of ground water levels and pressures, Bulletin No. 36, U.S. Army Engineers Waterways Experiment Station, Vicksburg, Mississippi.

Reades, D.W., Poland, R.J., Kelly, G., and King, S. (1987), discussion of "Predicting Hydraulic Conductivity of Two Prototype Clay Liners," by Day, S.R., and Daniel, D.E., J. Geotech. Engrg., ASCE, Vol. 113, No. 7, July 1987, pp. 809-813.

Workman, J.P. and Oakley, R.E. (1987), discussion of "Predicting Hydraulic Conductivity of Two Prototype Clay Liners," by Day, S.R. and Daniel, D.E., J. Geotech. Engrg., ASCE, Vol. 113, No. 7, July 1987, pp. 814-816.

ATTENUATING MATERIALS IN COMPOSITE LINERS

Jeffrey C. Evans, M.ASCE[1]
Yasodha Sambasivam[2]
Stephen Zarlinski[3]

ABSTRACT: Liners for hazardous and solid waste landfills are presently designed to minimize advective contaminant transport. The concept of composite liners designed with attenuating materials to minimize both advective and diffusive contaminant transport is explored. The mechanisms of diffusive transport and retardation are briefly reviewed. The properties of four materials (bentonite, zeolite, organically modified clay and fly ash) are examined in light of their expected performance in composite liner systems. It is concluded that liner systems should be designed with respect to the rate of contaminant transport. The technology for the necessary design calculations and optimization of liner performance is available. Studies by other researchers have been employed to support the concept of composite liners. It has been shown that the composite liners can be employed to reduce the rate of diffusive transport and to increase the time required for contaminant breakthrough.

INTRODUCTION

Hazardous and solid waste disposal facilities employ liner systems to impede the migration of contaminants into the environment. Hazardous waste disposal facilities in the United States typically employ liner systems which include two geosynthetic liners overlying natural or compacted clay.[1] Solid waste disposal facilities may be constructed with single or double liner systems depending upon state requirements. The "design philosophy" common to both is to design the liner system to minimize the rate of contaminant transport out of the containment facility by minimizing the hydraulic (advective) transport of contaminants through the liner. Designers recognize the potential for imperfections in the barrier layers. As a result, they incorporate leachate collection systems beneath the primary and secondary barriers to capture contaminants that may escape through the barrier layers. As part of the design process, the compatibility of the liner materials with the retained contaminants is also evaluated. Over the last decade research and design efforts have focused

---

[1]Assoc. Prof., Dept of Civil Eng., Bucknell University, Lewisburg, PA 17837
[2]Research Assoc., Dept of Civil Eng., Bucknell University, Lewisburg, PA 17837
[3]Laboratory Manager, Geosyntec Consultants, Norcross, GA 30093

upon liner materials which do not degrade in the presence of the retained contaminants. Present design philosophy reflects the uncertainty in the long-term effectiveness of liner systems by incorporating redundancy in the barrier layers through the concurrent use of both geosynthetic and clay materials.

Mechanisms of contaminant transport through barriers include migration in fluid due to hydraulic gradients (advection) and molecular diffusion in response to chemical concentration gradients. For liners designed to contain organic molecules of low molecular weight and constructed of relatively thin geomembranes, the rate of contaminant migration due to molecular diffusion may be significant. Molecular diffusion may be liquid phase diffusion through a saturated barrier layer or partitioning from the liquid phase may occur and result in gas phase diffusion. For example, a 30 mil geosynthetic liner was tested according to ASTM D1814-55 Inverted Cup Method and it was found that the rate of diffusion of the low octane gasoline was equal to approximately 3,000 gallons per acre in 24 hours due to molecular diffusion alone.[2] Although gasoline is not stored in geomembrane lined landfills, the tremendous rate at which low molecular weight organic contaminants (such as the volatile organics identified in the US EPA's priority pollutant list) can diffuse through geosynthetic liners is illustrated by this data.

The rate of contaminant migration is altered by a variety of factors including hydrolysis, sorption, cosolvation, ionization, precipitation and biotransformation.[3] Thus, in order to design a liner that minimizes the rate of contaminant migration into the environment, it is necessary to minimize the hydraulic conductivity and rate of diffusion and to maximize the retardation. Composite liners are suggested for this purpose. Composite liners are made of materials that minimize both the hydraulic conductivity and rate of diffusion and maximize the retardation through increased adsorptive capacity.

Several materials that have been proposed for use in composite liner systems to reduce the rate of diffusion and increase the attenuation capacity of the system are described in this paper. Through additional adsorptive capacity, the rate of contaminant migration can be reduced and the time for chemical breakthrough can be increased. The composite liner materials described in this paper include bentonite, chabazite (suitable for the inorganic components of the leachate) and organically modified clays and high carbon fly ash (suitable for organic constituents of the leachate).

This paper was prepared to focus the attention of landfill owners, operators, designers and regulators upon an aspect of liner design which has heretofore received little attention; that liner systems should be designed with respect to the rate of contaminant transport through the system. Further, the technology exists to perform the necessary design calculations and to optimize the liner performance by using composite liner systems. The results of a number of studies by other researchers have been examined and used to support the concept of composite liners. It has been shown that contaminant transport through conventional liner systems may be significant and that the technology exists to reduce the rate of diffusive transport. The time required for contaminant breakthrough can be greatly increased and the rate of diffusive transport can be decreased through the use of composite barriers.

PREVIOUS STUDIES

The application of naturally occurring materials with high adsorptive capacities to attenuate contaminant migration is well documented.[4] Many earlier landfills were designed as attenuating landfills, i.e., they were designed assuming leakage at a low rate on the basis that the undesirable constituents within the landfill leachate are attenuated by the underlying soils through adsorption, biological uptake, cation and anion exchange, dilution and precipitation.[5] This is also the fundamental principal behind the widespread use of septic tank-leach field systems. Considerable research has been done to describe the structure and the ion exchange capacity of a wide variety of naturally occurring clay materials.[6]

Recent research has contributed to our understanding of the factors influencing the hydraulic conductivity of clays in both the laboratory and field.[7,8,9,10] The transport of contaminants through geosynthetic liners has likewise been the subject of research and development.[11] It has long been recognized that impermeability is a myth.[12] Imperfections (e.g. pinholes, improperly sealed seams, tears as a result of construction) that occur in liner systems are accounted for in design through liner redundancy. Considerable research has been done in the area of compatibility between membranes and the contained waste.[13,14] These studies point to the well established notion that focuses on liners as barriers with low hydraulic conductivity. Finally, the concept of designing liner systems that release contaminants only by molecular diffusion has been examined.[15]

The previous studies have contributed to our ability to design liners with minimal hydraulic transport across the barrier layers. This paper presents concepts of composite liner systems which will also reduce the rate of diffusive transport across the liner.

CONTAMINANT MIGRATION THROUGH LINERS

The migration of contaminants by advection and molecular diffusion is well documented.[16] As the hydraulic conductivity decreases, the contaminant transport due to molecular diffusion becomes increasingly important. To properly characterize the contaminant transport through earthen barriers, both advective and diffusive transport must be considered. An excellent paper provides details of the governing equations, dimensionless charts and design methodology for contaminant transport through earthen barriers.[17] The purpose of this paper is to focus on the diffusive transport component and to describe materials that, when used in composite liner systems, will reduce the rate of diffusive transport. The limited theoretical description presented in this paper is not meant to be a comprehensive review of contaminant transport but is presented to set the background for the data presented later in the paper. For design of barrier systems, both advective and diffusive transport must be considered.

For the purposes of examining composite barriers to minimize diffusion, it is assumed that the seepage velocity approaches zero and the molecular diffusion rates are characterized by Fick's second law which describes one dimensional diffusion of solutes in soil as:

$$\partial c/\partial t = (D^*/R)(\partial^2 c/\partial x^2) \quad (1)$$

where:

D^* = effective diffusion coefficient
R = dimensionless retardation factor
c = concentration of a solute, and
t = time
x = direction of diffusion.

Thus the driving force (or gradient) for molecular diffusion is the chemical potential. Researchers have evaluated the diffusion coefficient for field conditions and compared these to laboratory values.[18] The effective diffusion coefficient which accounts for many of the real-world conditions including the tortuosity of the flow, temperature variations, and reactions which occur in the subsurface.[19] The apparent diffusion coefficient may be defined as D^*/R. This apparent diffusion coefficient includes retardation effects such as those due to adsorption.

Where the contaminant partitioning from the water phase to the solid phase can be described by the partition coefficient, the retardation can be related to the partition coefficient. For a fast reversible adsorption with a linear isotherm the partition coefficient can be replaced by the distribution coefficient. Retardation is then a function of the bulk density ρ, porosity η, and distribution coefficient K_d, as:

$$R = 1 + (\rho/\eta)K_d \quad (2)$$

where: R is the dimensionless retardation factor. As shown by equation (1), increasing the retardation factor, decreases the initial rate of diffusive transport. In composite liners, this can be accomplished using materials that demonstrate higher distribution coefficients. Additional discussion of diffusive transport in porous materials is found elsewhere.[19]

It is important to recognize that contaminants may migrate slower than the seepage velocity of the transporting solution. This attenuation can be due to adsorption, ion exchange, precipitation and biotransformation. These mechanisms which reduce the rate of transport can be incorporated into the design of composite liners to reduce the rate of contaminant transport across the barrier.

COMPOSITE LINER CONCEPTS

Shown on Figure 1 is a liner system incorporating two geomembrane barrier layers overlying a compacted clay layer. This liner system is typical of that required for containment of hazardous waste. Under the present system the geosynthetic and clay liners are all acting as hydraulic barriers. In a sense, the present system is already a composite liner using different materials in a redundant hydraulic barrier system. The leachate collection system in the interlayer between the primary and secondary geomembrane liner is designed to collect aqueous wastes which have migrated through the primary barrier layer.

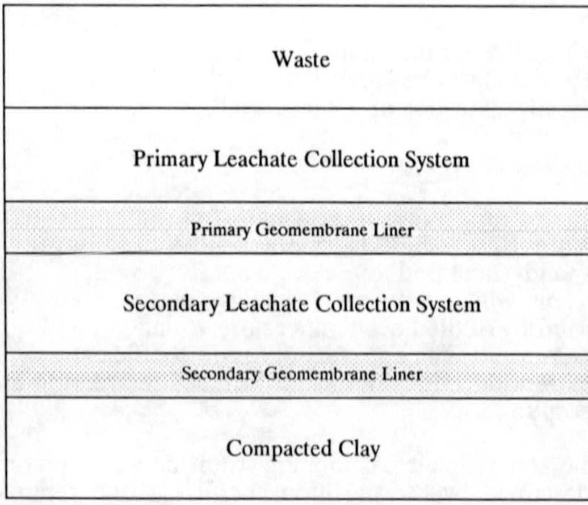

Figure 1. Typical Hazardous Waste Liner System

Shown on Figure 2 is a conceptual composite liner system. This conceptual composite liner system incorporates the present barrier and leachate collection layers with additional modification to increase the attenuating capacity of the liner system. Depending upon the design criteria, the attenuating materials can be admixed with the underlying compacted or undisturbed clay. Based upon present regulatory requirements for hazardous waste sites, the attenuating layers would likely be in addition to the compacted clay. For solid waste landfills, the attenuating layers could be in lieu of an equivalent thickness of compacted or undisturbed clay. Attenuating materials could include calcium or sodium bentonite and/or zeolites for inorganics. Materials to attenuate organics could include organically modified clays and/or fly ash. Organically modified clays are most effective in retaining higher molecular weight organics whereas the carbon in the fly ash is effective for low molecular weight organics.

The thickness of the attenuating layers will depend upon the hydraulic conductivity, hydraulic head, chemical potential, effective diffusivity, retardation factor and design requirements defining allowable breakthrough time and steady state transport rates. The method of calculation can be based upon the analytical solution to the advective-dispersive equation for solute transport such as described elsewhere.[17]

The adsorbing barrier layers may be placed as separate layers, admixed with the clay layers, or attached to the geomembranes. For example, admixed soil liners are frequently constructed on native soil with a bentonite admixture to reduce hydraulic conductivity. Admixing attenuating materials can result in material of relatively low hydraulic conductivity ($< 1x10^{-7}$ cm/sec)[20,21] thereby controlling advective transport and providing assurance that diffusive transport is a significant component of the overall transport. Alternatively, attaching clays to geomembrane barriers is a well developed manufacturing technique. Paraseal, of Paramount Technical Products, Inc., is one such example.[22]

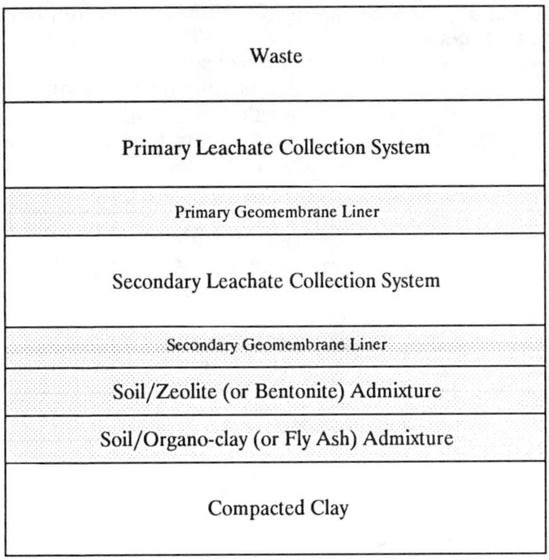

Figure 2. Conceptual Composite Liner System

The proposed composite liner would be more effective in minimizing the rate of contaminant transport in comparison to liners presently used. The improved effectiveness during transport is accomplished through adsorption and cation exchange of inorganics within the zeolite and/or bentonite and natural clays, and adsorption of organics by the organically modified clays and/or carbon rich fly ash. The adsorptive layers should be located on the underside of the secondary geomembrane liner. In this way, the geomembrane would function as the primary barrier against contaminant transport by acting as a hydraulic barrier. This reduces the contaminant loading to the adsorptive layers and increases the time during which they will be effective. Contaminants which do migrate through the geomembrane liners (either through diffusion or advection through liner defects) would be adsorbed by the underlying attenuating materials. It is cautioned that further study may show that, although high carbon fly ash is effective for organic contaminants, the heavy metals leached may exceed the attenuation capacity of the underlying clays.

The remainder of the paper discusses the individual components in detail and provides test results to support the proposed concept of the composite liner to minimize contaminant transport. It is interesting to note that the use of these materials individually as adsorption layers is not a new concept.

Zeolite, carbon, and organically modified clays have all been used in waste-water treatment processes.[23]

COMPOSITE LINER MATERIALS

Bentonite. Bentonites are clays which contain montmorillonite (a smectite) as the major clay constituent. Calcium and sodium bentonite are the two major types of bentonites. The distinction is based upon the ion that dominates the cation exchange reaction (typically 35-99% of the exchangeable cations). The cation exchange capacities of calcium and sodium bentonites are approximately 80 and 150 milliequivalents per 100 grams of dry soil, respectively.[6] Calcium bentonites are commonly found and mined in areas of Mississippi and Alabama, whereas sodium bentonites are found and mined in Wyoming and South Dakota.

Calcium bentonite swells less than sodium bentonite upon wetting with water. This reduced swelling compared to sodium bentonite may not be a detrimental characteristic since the primary function of the bentonite in the composite liner system is to serve as an ion exchange layer rather than a hydraulic barrier. The divalent calcium is, however, less readily exchangeable than the monovalent sodium. The exchangeable ions (calcium and sodium) in the clay readily exchange with heavy metals such as lead (Pb), copper (Cu), nickel (Ni), and lead (Pb).

Zeolite. Zeolites are crystalline hydrated 3-dimensional alumino-silicates with a caged structure through which the exchangeable cations are adsorbed. Zeolites have a cation exchange capacity of 250 milliequivalents per 100 grams with the major exchangeable cations being sodium and calcium. In the caged structure, the zeolite acts like a sieve when permeated by water. Metals that pass through the sieve are trapped via ion exchange. Zeolite can complement bentonite in the composite lining system to enhance adsorption of metals and other cations.

The five major naturally occurring zeolites are analcite, chabazite, clinoptilolite, erionite, and mordenite. Zeolites are presently used to purify water contaminated with radioactive ions, remove ammonium from the effluent of waste-water treatment plants, purify water in fish tanks, control odor in kitty-litter, and adsorb metals from industrial waste waters.

Arsenic, lead, and cadmium are preferentially adsorbed by zeolites. A study by the Montana College of Mineral Science and Technology demonstrated that chabazite was an excellent adsorption medium for metals in groundwater. The structure and behavior of zeolites have been studied extensively and are well established.[24]

In the composite liner system, zeolite could be blended with bentonite or placed in a separate layer. Placement of the zeolite as a separate layer above the bentonite would prevent exchange reactions of the calcium from the bentonite with the sodium from the zeolite. It should also be noted that sodium and calcium will be released in the cation exchange process and any environmental consequences of this release should be considered.

Organically Modified Clays. Organically modified clays were initially developed about 40 years ago when Jordan published the first original work on the subject.[25,26] Organo-clays have been used in a number of environmental applications as described by Alther et al.[27], and Evans et al.[23] Organically modified clays are naturally occurring clays in which a portion of the inorganic exchangeable cations are exchanged with a suitable organic cation, typically a quaternary ammonium. Quaternary ammonium is a type of organic nitrogen compound in which the molecular structure includes a central nitrogen atom joined to four organic groups along with an acid radical. The reactions which exchange the inorganic cations on the surface of the clay with the organic cations to form the organically modified clay include adsorption, ion exchange, and intercalation. Differences in organically modified clays result from differences in the base clay (i.e montmorillonite, attapulgite) and differences in the organic cation used in the exchange.

Organically modified clays adsorb organic molecules as a result of two primary phenomena.[28,29] These phenomena are adsorbate interactions and adsorbate-solvent interactions. The adsorption capacity of an organically modified clay is therefore dependant upon both the properties of the material used to convert the clay to its organophilic nature and the properties of the medium such as temperature, pH, and type of solvent. Considerable insight into the behavior of organically modified clays has been provided by Boyd et al.[30], Mortland et al.[31], and Woolf, et al.[32] Boyd et al. and Mortland, et al. have demonstrated that clays modified with strongly hydrophobic organic molecules tend to be excellent adsorbents of chlorophenols, but are less effective for non-chlorinated phenols. In essence, the nature of the organically modified clay will control its ability to preferentially adsorb specific organic waste constituents.

Three organically modified clays were tested by Woolf et al. to ascertain their efficiency when compared to activated carbon in the removal of a variety of organic compounds with water; the results of the study were encouraging. Boyd et al. have also shown that as the hydrophobicity of the adsorbate increases, the adsorptive capacity of organically modified clay increases. They also demonstrated that the reaction of primary amines (an amine is a compound derived from ammonia by replacement of hydrogen by one or more univalent hydrocarbon radicals) with other molecules is pH dependant whereas this is not the case for quaternary amines, making them more suitable adsorbent for the decontamination of water.

The process of sorbing organics onto organically modified clays from water is a partitioning process. Boyd et al. describes this process such that the quaternary ammonium bound to the organically modified clay surface acts as a solubilizing or partitioning media to remove organic molecules from water. This is functionally and conceptually similar to a bulk organic solvent such as hexane or octanol removing (extracting) organics from water. In their studies, they use hexadecal, trimethyl ammonium chloride ions (which are 16 carbons long) to remove non-ionic, organic contaminants such as benzene, dichlorobenzene, and perchloroethylenane in water with excellent success. The evidence points to the effectiveness of organically modified clays for the removal of toxic compounds from waste water. In the composite liner case, these organically modified clays remove organic contaminants from the leachate.

Fly Ash. A description of the fly ash properties best suited for landfill liner applications is found in Mott and Weber.[33] These authors have demonstrated that the molecular diffusion of organics results in solute breakthrough in a relatively short time period through slurry trench cutoff walls. The addition of high carbon fly ash within the barrier results in a substantial delay in breakthrough as will be subsequently discussed in detail. It is also important to note that high carbon fly ash shows an adsorptive preference for low molecular weight organics in contrast to organically modified clays which preferentially adsorb high molecular weight organics.

COMPOSITE LINER PERFORMANCE

Natural Clay. The effect of sorption on diffusion rates has been examined for natural clay beneath a hazardous waste landfill.[18] Based upon a comprehensive study employing both field and laboratory data, the effective diffusion coefficient was from 0.2 to 0.02 less the theoretical diffusion coefficient for chloride. The reduction was attributed to the sorption and molecular weight effects. The authors point out that diffusion of organics such as benzene will be further retarded compared with the chloride. However, given the high toxicity of organics such as benzene, it is appropriate to calculate the breakthrough time for small values of C/C_o (they suggest 0.001). Thus, for their field data from a hazardous waste landfill, the time for benzene to breakthrough a one meter thick clay layer (at $C/C_o = 0.001$) was calculated to be 70 years.

The importance of diffusion for landfill liners through clay was demonstrated by the calculation of Johnston, et al. of the release rate for a contaminant diffusing through a 1 meter thick liner.[18] Using the data developed in their study, and assuming a landfill area of 10^4 m^2 and a concentration of 1 g/L in the landfill, the authors compute a steady-state transport rate of 19 Kg/year. Assuming the contaminant to be benzene, this transport rate has the potential to contaminate 3.8 billion liters of water at the U.S. EPA drinking water limit of 0.005 mg/L.

The authors[18] go on to argue that diffusion continues even with double liners and leachate collection systems (similar to Figure 1). In fact, the conclude that the diffusion rate through a double liner system is similar to a single layer barrier of equal thickness.[18]

It is concluded that natural clays offer sorption capacity for both organics and inorganics and are superior to synthetic liners in this respect. Composite liner systems offer increased sorptive capacity which result in increased breakthrough times (and prolong the time required to reach steady state mass flux across the barrier).

Fly Ash. In a recent study, the effectiveness of a composite soil-bentonite and fly ash cutoff wall as a contaminant transport barrier was investigated.[34] High carbon fly ash was employed to reduce diffusive transport through the barrier. As previously noted, all other conditions being equal, the rate of diffusion through a barrier layer is controlled by the effective diffusion coefficient; thus composite barriers with lower effective diffusion coefficients will serve as better contaminant barriers than their conventional counterparts.

The investigators conducted both quasi-steady-state and transient diffusion experiments to measure the effective diffusion coefficient for selected solutes through conventional soil-bentonite and through soil-bentonite and fly ash composite barriers. Since the diffusive transport rates are hindered by sorption in the media, the apparent diffusion coefficient (which incorporates both the effective diffusion coefficient and the retardation) was related to the sorption capacity of the fly ash. For these experiments, the researchers found that the composite barrier modeling required both non-equilibrium and non-linearity in the sorption term.

Using the results of the above experiments, the researchers[34] simulated a soil-bentonite cutoff wall 1 meter thick. It was assumed that a volatile organic solute was at 1 mg/L on the interior and 0.0 mg/L on the outside of the barrier. The results of the modeling are shown on Figures 3 and 4 for the volatile organic solute. These results are for the two limiting cases where no sorptive capacity is assumed for soil-bentonite and equilibrium sorptive capacity is assumed for the composite barrier. For the total contaminant migration, it is necessary to multiply the results on Figure 3 by the actual interior concentration and the area of the barrier wall. For a wall containing a square site of 10 acres and extending 50 feet below the water table, the total contaminant migration would be about 600 Kg in the first 12 years with a conventional cutoff wall. In contrast, the total transport is essentially zero for the composite wall for the same period. It would require about 100 years for the sorption capacity of the composite barrier to be exhausted and for the transport rate to be equal to that of the conventional cutoff wall.

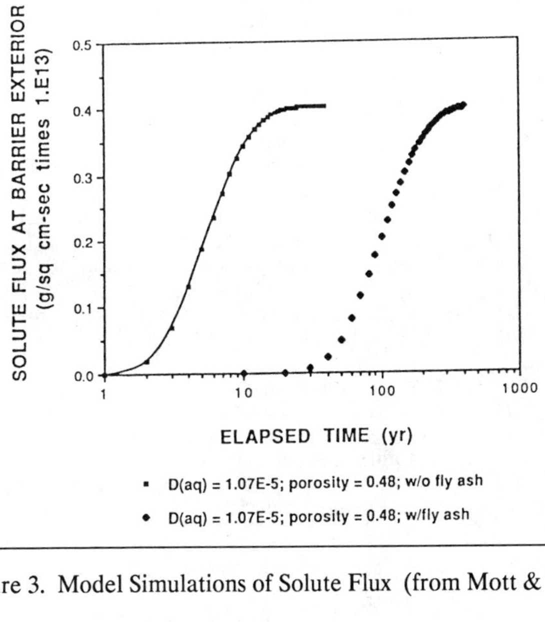

Figure 3. Model Simulations of Solute Flux (from Mott & Weber, 1989)

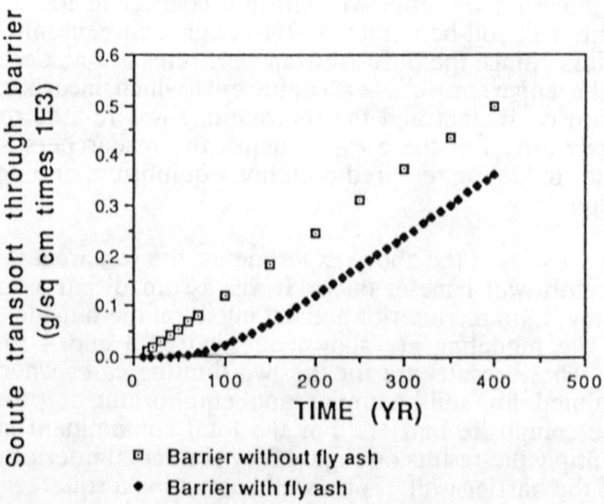

Figure 4. Model Simulations of Solute Transport (from Mott & Weber, 1989)

Bentonite. The effect of bentonite in retarding contaminant migration was examined as part of a study of diffusion through barrier materials.[35] The study employed silica sand/bentonite mixtures and the reactive solute strontium 85. Shown on Figure 5 are the relationships between the retardation factor R, the distribution coefficient K_d, and the bulk density/porosity ratio. As the bentonite content increases, the overall sorptive capacity increases and thus, as shown, the measured values of K_d increase. With all other factors being equal, increasing the bentonite content, and thus K_d, would result in a increasing retardation factor. However, increasing the bentonite content also generally resulted in a decreasing density/porosity ratio. As a result, the retardation factor is optimized at a bentonite content between 5 and 10 percent by weight. The resulting diffusion coefficients are shown on Figure 6. The figure also shows that the optimum bentonite content is between 5 and 10 percent by weight. It is interesting to note that the bentonite content of 5 to 10 percent found to optimize the diffusion coefficient is similar to the bentonite content found to minimize hydraulic conductivity in admixed soil liners.

The results of this study provide additional insight into the composite liner approach. First, the addition of bentonite to a liner system adds adsorptive capacity and results in a reduced apparent diffusion coefficient. Second, large increases in bentonite content may not always be beneficial. As with a number of natural systems, there may exist an optimum which needs to be assessed through experimentation and analysis.

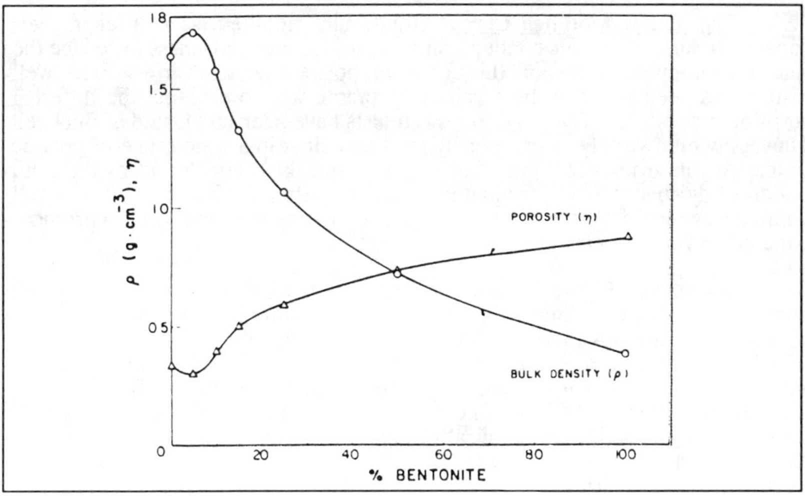

Figure 5. Strontium-85 Coefficients, Bulk Density to Porosity Ratios and Retardation Factors versus Bentonite Content (from Gillham, et al., 1984)

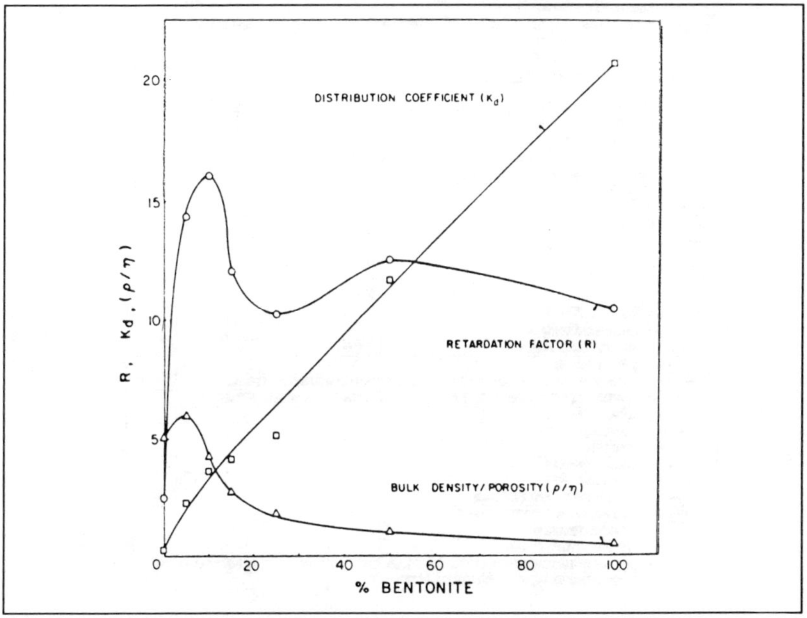

Figure 6. Strontium-85 Diffusion Coefficients: Experimental & Calculated versus Bentonite Content (from Gillham, et al., 1984)

Organically Modified Clays. Organically modified clays differ in their capacity to adsorb organic contaminants. Thus, their effectiveness to reduce the rate of contaminant transport through a composite liner also varies. Free swell testing has been found to be a relatively simple way to examine the differing sorption capabilities. Over 100 free swell tests have been conducted at Bucknell University on a variety of organically modified clays in a wide range of organic fluids. In these tests, 2.5 grams of clay was sprinkled into 50 ml of the fluid under study in a 100 ml graduated cylinder. After 24 hours the free swell volume was recorded. Occasional low energy stirring was employed to promote contact between the clay and fluid.

Shown on Figure 7 are the results of free swell tests using the concentrated organic fluid hexane sorbed by bentonite, attapulgite, and a number of organically modified clays. The free swell volumes ranged as high as 47 ml for the organically modified clays and were 5 and 7 ml for the untreated bentonite and attapulgite, respectively. In essence, 2.5 g of the organophilic clay Tixogel EZ-100 sorbed 47 ml of hexane. Although the sorption capacity would undoubtedly be less for an organically modified clay in a composite liner system, the significant sorption capacity of these clays make them especially well suited as sorbents of organics.

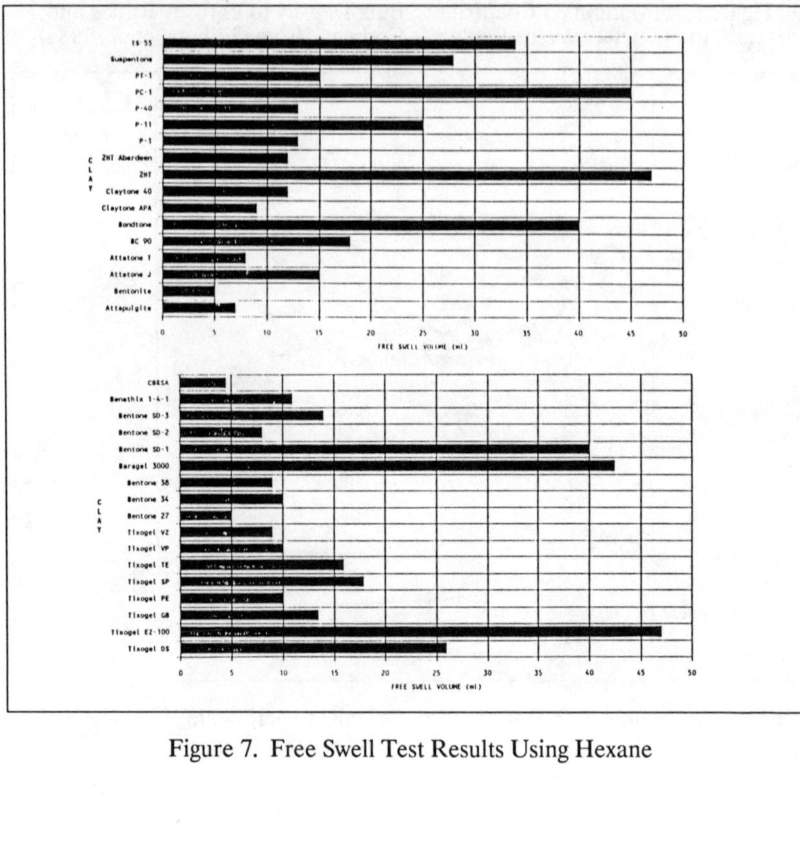

Figure 7. Free Swell Test Results Using Hexane

The effectiveness of organically modified clays in removing organics from aqueous solutions[36] is shown on Figure 8. The figure presents the influent and effluent concentrations of oil in an aqueous solution "treated" with an organically modified clay. The oil is preferentially sorbed to the organically modified clay reducing the contaminant levels in the effluent. Although this data represents a waste treatment application it is conceptually similar to the composite liner application.

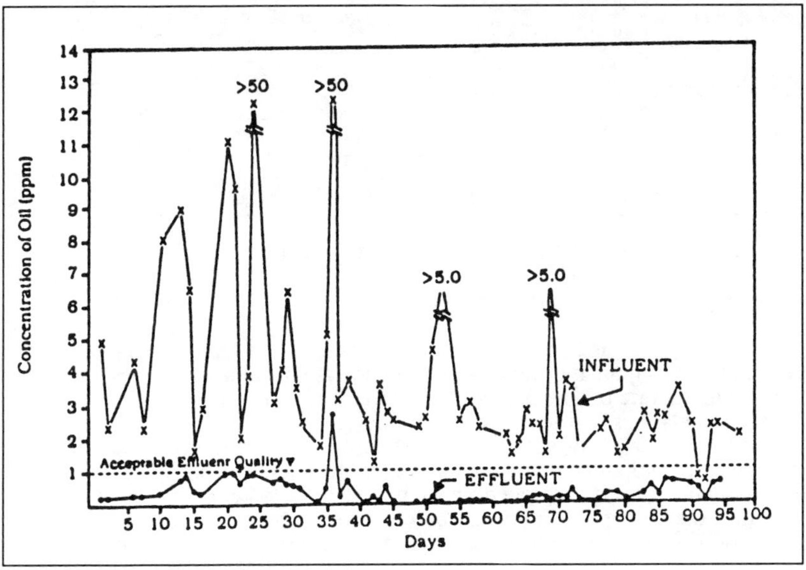

Figure 8. Influent and Effluent Oil Concentrations (from Electrum, Inc.[36])

Zeolite. The adsorption characteristics of chabazite were examined in triaxial permeability tests as shown on Figure 9. Tests on a silty clay and a silty clay chabazite composite were conducted using a solution of lead at 10 mg/l in distilled water. Note the retardation of the lead in the composite liner as compared to the compacted silty clay. As a basis of comparison the results for copper are also shown. First, it is observed that the copper reaches breakthrough faster (as measured by pore volume displacement) than the lead for both liners. This reveals the difference in retardation factors between copper and lead. Further, in the chabazite composite barrier, lead is retarded compared with the silty clay barrier. Chabazite preferentially adsorbs lead as compared to the silty clay whereas the same preference is not shown for the copper. These results emphasize the importance of designing composite liners to sorb contaminants of interest. That is, the interactions between composite materials and specific contaminants must be examined to optimize the composite liner performance.

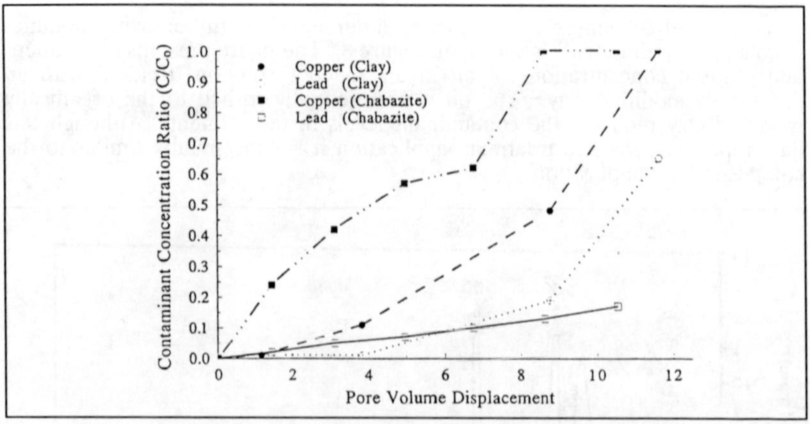

Figure 9. Permeation of Chabazite and Silty Clay

SUMMARY AND RECOMMENDATIONS FOR FURTHER STUDY

This paper attempts to focus the attention of landfill owners, operators, and designers upon the concept that liner systems should be designed with respect to the rate of contaminant transport. The technology is available for the design calculations and optimization of liner performance using composite liner systems. A number of studies by other researchers have been examined and employed to support the concept of composite liners. It has been shown that contaminant transport through conventional liner systems may be significant and that the composite liners can be employed to reduce the rate of diffusive transport and to increase the time required for contaminant breakthrough.

It is recognized that the sorptive capacity of the materials discussed is finite. However, as the studies using a soil-bentonite and fly ash composite described in this paper demonstrate, the breakthrough time may be delayed by decades by increasing the adsorptive capacity of the liner system. Additional studies of a similar nature are required for each of the composite materials described in this paper to quantify the affect upon breakthrough time and steady-state transport rates.

Additional study is also required to optimize the design of composite liner systems in order to maximize breakthrough time and minimize diffusive transport rates for a cost-effective barrier system. It is necessary to correlate the sorption capacity as measured by adsorption isotherms to liner performance. It is also necessary to investigate the sequence of layer placement which yields the largest breakthrough time and lowest diffusive transport rate. Finally, it is necessary to evaluate the effect of the release of exchangeable ions from the composite materials upon the environment. As with any new approach, it is necessary to evaluate secondary effects such as slope stability of the composite liner. As the data described in this study indicate, the performance of the composite materials is contaminant specific. Therefore, the optimum design for one waste facility may not be the same as that for another.

ACKNOWLEDGEMENTS

The authors appreciate the detailed comments of the ASCE reviewers of this paper. Their efforts have contributed significantly to this paper. Also appreciated are the review comments of George Alther of Bentec, Inc.

REFERENCES

[1]The Comprehensive Environmental Responsibility, Compensation and Liability Act, (Superfund), 42 USC 9601 et seq., 1980.

[2]Seaman Corporation, XR5 Geomembrane literature, Wooster, Ohio

[3]Johnson, R. L., Palmer, C. D., and Fish, W., "Subsurface Chemical Processes", *Transport and Fate of Contaminants in the Subsurface*, EPA/625/4-89/019, Sept., 1989.

[4]Bagchi, A. Design, Construction & Monitoring of Sanitary Landfill, John Wiley & Sons, New York, 1990

[5]Bagchi, A. Design of Attenuating Landfills, *Journal of Environmental Engineering Div., ASCE*, 110, No. EE-6, 1211-1212.

[6]Grim, R. E. Clay Mineralogy. McGraw-Hill Book Company, Inc., New York, 1968.

[7]Daniel, D. E.., S.J. Trautwein, S. S. Boynton, and D. E.. Foreman. 1984. "Permeability Testing with Flexible Wall Permeameters". Geological Testing Journal. 7(3):113-122.

[8]Dunn, R. J., and J. K. Mitchell. 1984. " Fluid Conductivity Testing of Fine Grained Soils". Journal of Geotechnical Engineering. 110(11):1648-1665.

[9]Evans, J. C. and Fang, H. Y., "Triaxial Permeability and Strength Testing of Contaminated Soils," *Advanced Triaxial Testing of Soil and Rock*, ASTM STP 977, American Society for Testing and Materials, Philadelphia, 1988. pp. 387-404.

[10]Olson, R. E., and D. E.. Daniel. 1979. "Field and Laboratory Measurements of the Permeability of Saturated and Partially Saturated Fine-Grained Soils". In: ASTM Symposium on Permeability and Groundwater Contaminant Transport, Philadelphia, Pennsylvania. 67 pp.

[11]Haxo, H. E. Jr., Miedema, J. A., and Nelson, N. A., "Permeability of Polymeric Membrane Liner Systems", *Proceedings of the International Conference on Geomembranes*, Denver, CO, 1984, pp.151-156.

[12]Giroud, J. P., Impermeability: The Myth and a Rational Approach", *Proceedings of the International Conference on Geomembranes*, Denver, CO, 1984, pp.157-162.

[13]Verschoor, K., Britton, L., and Thomas, R., "Chemical Compatibility Testing of Geosynthetics to be Used as Containment Barriers in Hazardous Waste Management," Hazardous Waste and Hazardous Materials, 1988, pp. 205-209.

[14]Haxo, H. E., Haxo, R. S., Nelson, N. A., Haxo, P. D., White, R. M., Dakessian, S., and Fong, M. A., "Liner Materials for Hazardous and Toxic Waste and Municipal Solid Waste Leachate," Noyes Publications, 1985.

[15]Daniel, D. E. and C. D. Shackelford, Disposal Barriers that Release Contaminants Only by Molecular Diffusion, Proceedings of the Oak Ridge Model Conference, Oak Ridge National Laboratory, Oak Ridge, Tenn., Vol. I, No. 3, 1987, pp. 315-332.

[16]Freeze, R. A. and Cherry, J. A. Groundwater, Prentice-Hall, Englewood Cliffs, N. J., 1979, 604 pp.

[17]Shackelford, C. D., "Transit-time Design of Earthen Barriers," *Engineering Geology*, 1990, Vol. 29, pp. 79-94.

[18]Johnson, R. L., Cherry, J. A., Pankow, J. F., "Diffusive Contaminant Transport In Natural Clay: A Field Example and Implications for Clay-Lined Waste Disposal Sites", Environ. Sci. Technol., Vol. 23, No. 3, 1989, pp. 340-349.

[19]Shackelford, C. D., "Diffusion of Contaminants Through Waste Containment Barriers," *Transportation Research Record* 1219, Geotechnical Engineering 1989, Transportation Research Board, National Research Council, pp. 169-182.

[20]Smalstig, E., "Comparison Study of Different Clay Liners used as Barriers to Contaminant Transport," unpublished student independent laboratory report, Bucknell University, May, 1990.

[21]Mead, T., Composite Slurry Walls with Organo-clays," unpublished student independent laboratory report, Bucknell University, May, 1990.

[22]Paraseal, Paramount Technical Products, Inc., Bloomington, MN, 55431

[23]Evans, J. C., Pancoski, S. E., "Organic Waste Treatment with Organically Modified Clays" *Proceedings of the 3rd International Conference on New Frontiers for Hazardous Waste Management*, Pittsburgh, PA, September 10-13, 1989, EPA/600/9-89/072, pp. 48-57.

[24]Mumpton, F. A., Mineralogy and Geology of Zeolites, Mineralogical Society of America, Short Course Notes, Southern Printing Co., Blacksburg, Va., Vol. 4, 1977, 233 pp.

[25]Jordan, J. W., "Alteration of the Properties of Bentonite by Reaction with Amines," Mineralogical Magazine and Journal of the Mineralogical Society, Vol. 28, No. 205, June 1949, pp. 598-605.

[26]Jordan, J. W., "Organophilic Bentonites: Swelling in Organic Liquids," The Journal of Physical & Colloid Chemistry, Vol. 53, No. 2, February 1949, pp. 294-306.

[27]Alther, G. R.; Evans, J. C., and Pancoski, S. E., "Organically Modified Clays for Stabilization of Organic Hazardous Wastes," Superfund '88, *Proceedings of the Ninth National Conference, Hazardous Materials Control Research Institute*, November 1988, pp. 440-445.

[28]Chiou, C. T., Porter, P. E., and Schmedding, D. W., "Partition equilibria of Nonionic Organic compounds between Soil Organic Matter and Water," Environmental Science and Technology, 17, 1983, pp. 227-231.(check this)

[29]Chiou, C. T., Peters, L. J., and Freed, V. H., "A Physical Concept of Soil-Water Equilibria for Nonionic Organic Compounds", Science, 213, pp. 684-685.

[30]Boyd, S. A., Lee, J-F, Mortland, M. M. "Attenuating Organic Contaminant Mobility by Soil Modification," Nature, Vol. 333, May, 1988, pp. 345-347.

[31]Mortland, M. M.; Shaobai, S.; and Boyd, S. A., "Clay-Organic Complexes as Adsorbents for Phenol and Chlorophenols," Clays and Clay Minerals, Vol. 34, No. 5, pp. 581-585, 1986.

[32]Wolf, T. A., Demirel, T., and Bauman, R. E., "Adsorption of Organic Pollutants on Montmorillonite Treated with Amines," Journal of the Water Pollution Control Federation, 58, 1986, pp. 68-76.

[33]Mott, H. V., Weber, W. J., "Diffusive Transport and Attenuation of Organic Leachates in Cut-Off Wall Backfill Mixes," Presented at the Twelfth Annual Madison Waste Conference, September 20-21, 1989, Department of Engineering Professional Development, University of Wisconsin-Madison.

[34]Mott, H. V. and Weber, W. J., "Solute Migration in Soil-Bentonite Containment Barriers," Superfund '89, *Proceedings of the Ninth National Conference, Hazardous Materials Control Research Institute*, November, 1989, pp. 526-533.

[35]Gillham, R. W., Robin, M. J. L., Dytynyshyn, D. J., and Johnston, H. M., "Diffusion of Nonreactive and Reactive Solutes through Fine-Grained Barrier Materials", Can. Geotech. J., 21,pp. 541-550, 1984.

[36]Electrum, Inc., "Organosorb Cleanup of Oily Steam Condensate in a Petrochemical Plant," Company literature, Fairfield, Kentucky

SUBJECT INDEX

Page number refers to first page of paper.

AUTHOR INDEX

Page number refers to first page of paper.